2nd Edition

AP° ENVIRONMENTAL SCIENCE
CRASH COURSE°

By Gayle N. Evans, M.Ed.
Updated by Amy Fassler, M.A.

Research & Education Association
www.rea.com

ABOUT REA

Founded in 1959, Research & Education Association (REA) is dedicated to publishing the finest and most effective educational materials—including study guides and test preps—for students of all ages.

Today, REA's wide-ranging catalog is a leading resource for students, teachers, and other professionals. Visit www.rea.com to see our complete catalog.

Research & Education Association
258 Prospect Plains Road
Cranbury, New Jersey 08512
Email: info@rea.com

AP® ENVIRONMENTAL SCIENCE CRASH COURSE, 2nd Edition

Library of Congress Control Number 2019949076

ISBN-13: 978-0-7386-1256-0
ISBN-10: 0-7386-1256-1

AP® Environmental Science Crash Course
TABLE OF CONTENTS

PART I INTRODUCTION

PART II LIFE ON EARTH

UNIT 1 | THE LIVING WORLD: ECOSYSTEMS

UNIT 2 | THE LIVING WORLD: BIODIVERSITY

HOW HUMANS USE AND CHANGE EARTH

ENVIRONMENTAL OVERVIEW

TEST-TAKING STRATEGIES AND PRACTICE

PRACTICE EXAM ....*www.rea.com/studycenter*

ABOUT OUR BOOK

REA's *AP® Environmental Science Crash Course* is designed for the last-minute studier or any AP® student who wants a quick refresher on the course. The *Crash Course* is based on the latest changes to the AP® Environmental Science course and exam and focuses only on the topics tested, so you can make the most of your study time.

Written by veteran AP® Environmental Science test experts, our *Crash Course* gives you a concise review of the major concepts and important topics tested on the exam.

- **Part I** gives you the **Keys for Success**, so you can tackle the exam with confidence. It also gives you a review of basic math and science concepts as well as the **Key Terms** that you absolutely, positively must know.

- **Part II** presents essential information about **Life on Earth** including geological concepts, organisms, and atmospheric conditions.

- **Part III** discusses how humans use and change the Earth—from pollution to energy resources to species extinction—concentrating on what you need to know for the exam.

- **Part IV** puts environmental science into context with other disciplines such as economics and human geography. This broader picture helps you respond to questions that involve issues (e.g., climate change) for which environmental science is a key, but not the only, part.

- **Part V** gives you specific **Test-Taking Strategies** to help you conquer the multiple-choice and free-response questions, along with AP®-style practice questions to prepare you for what you'll see on test day.

ABOUT OUR ONLINE PRACTICE TEST

How ready are you for the AP® Environmental Science exam? Find out by taking REA's online practice exam available at *www.rea.com/studycenter*. This test features automatic scoring, detailed explanations of all answers, and diagnostic score reporting that will help you identify your strengths and weaknesses so you'll be ready on exam day!

Whether you use this book throughout the school year or as a refresher in the final weeks before the exam, REA's *Crash Course* will show you how to study efficiently and strategically, so you can boost your score.

Good luck on your AP® Environmental Science exam!

ABOUT OUR AUTHORS

Gayle Evans, M.Ed., is a Lecturer and Science Master Teacher at the University of Florida's School of Teaching and Learning. Her career includes 14 years at Gainesville (Fla.) High School, where she taught a wide range of science courses including AP® Environmental Science and AP® Biology.

Ms. Evans earned her B.A. in biology from Mount Holyoke College in Massachusetts and her M.Ed. in secondary science education from the University of Florida in Gainesville. She holds National Board Certification in Biological Sciences for Adolescents and Young Adults.

Amy Fassler, M.A., teaches AP® Environmental Science at Marshfield (Wisc.) High School, where she has taught science for 24 years.

A former member of both the College Board's AP® Environmental Science Test Development Committee and the Instructional Design Team for the course, she has served as an AP® Table Leader, AP® Question Leader, and AP® Reader.

Ms. Fassler has also developed curriculum and run teacher workshops for the Howard Hughes Medical Institute.

A lifelong advocate for environmental causes, she coaches the Science Olympiad Team at Marshfield High.

Ms. Fassler holds a Master Educator License in environmental science, life science, and chemistry. She earned her B.A. in biology and chemistry from Black Hills State University and her M.A. in secondary science education from Viterbo University. Ms. Fassler's passion for environmental science education has inspired many young people to pursue careers in the field, or simply become more environmentally aware citizens.

ACKNOWLEDGMENTS

We would like to thank Larry B. Kling, Editorial Director, for his overall guidance; Pam Weston, Publisher, for setting the quality standards for production integrity and managing the publication to completion; John Cording, Technology Director, for coordinating the design and development of the REA Study Center; and Wayne Barr, Test Prep Project Manager, for editorial project management; and Jennifer Calhoun for file prep.

We would also like to extend our appreciation to Pamela Shlachtman for technically reviewing the manuscript; and Kathy Caratozzolo of Caragraphics for typesetting this edition.

PART I

INTRODUCTION

Keys for Success
on the AP® Environmental Science Exam

The AP® Environmental Science exam will test your ability to identify and analyze natural and human-made environmental problems. You will be asked to describe the risks associated with these problems and provide realistic solutions. Environmental science is interdisciplinary, so the exam will require you to explain connections from a broad range of topics including biology, chemistry, geography, and environmental studies.

When you sit for the AP® Environmental Science exam, you will find that it evaluates much more than your knowledge about environmental science concepts. In some ways, you are also being assessed on how well you handle taking the test itself. This *Crash Course* is based on a careful analysis of the most recent College Board *AP® Environmental Science Course and Exam Description (CED)*. Our book gives you the content, skills, and strategies to be successful on the exam—all in one compact volume.

1. **Understanding the AP® Environmental Science Scoring Scale**

    The exam is scored on a 5-point scale. The AP® exam administrators have determined general guidelines, where achieving a 3, 4, or 5 typically qualifies students to earn credit for an equivalent college course (check the specific policy of the college or university where you wish to enroll). Unlike high school classes where your work may be graded on a curve, or norm-referenced, AP® exams are criterion-referenced. This means that any student who meets the criteria, or score cutoff, will get a qualifying score. When the exams are read and scored each year, cut scores are set that determine a score of 5, 4, 3, 2, or 1. A composite score of 50% to 55% is generally the cutoff for a 3. Roughly 50% of students who take the AP® Environmental Science exam get a score of 3 or higher.

The AP® Environmental Science exam consists of 2 sections as outlined below. You are allowed to use a calculator for both sections of the exam. A list of approved calculators can be found on the College Board's AP® Central website.

Section 1: Multiple-Choice

- 80 questions
- 12–15 common option sets
- 65–68 classification sets and discrete items
- Exam Weight = 60%
- 90 minutes

Section 2: Free-Response

- 3 questions, 10 points each
- Question 1: Design an investigation
- Question 2: Analyze an environmental problem and propose a solution
- Question 3: Analyze an environmental problem and propose a solution using calculations
- Exam Weight = 40%
- 70 minutes

2. **Understanding the AP® Environmental Science Course Framework**

The AP® Environmental Science course framework provides a clear and detailed description of the requirements necessary for students to be successful on the exam. It includes two key components. The *Science Practices* describe what students should be able to do to show they understand the course concepts and the *Course Content* details the conceptual understanding on which students will be assessed. Before you begin your exam preparation we recommend that you review the *AP® Environmental Science Course and Exam Description*, which is freely available at the College Board's AP® Central website. You'll find a detailed description of the both the Science Practices and the Course Content.

The course content is broken down into nine units organized around Big Ideas, core principles that spiral throughout every unit.

Big Ideas

Big Idea 1: Energy Transfer (energy conversion behind all ecological processes)

Big Idea 2: Interactions Between Earth Systems (assess system changes over time and space)

Big Idea 3: Interactions Between Different Species and the Environment (impact of humans on the environment, both good and bad)

Big Idea 4: Sustainability (preserving the environment through a combination of conservation and development)

Here are the nine units that make up the course content, along with the exam weighting for each unit.

AP® Unit Breakdown

Unit	Exam Weighting
Unit 1: The Living World: Ecosystems	6%–8%
Unit 2: The Living World: Biodiversity	6%–8%
Unit 3: Populations	10%–15%
Unit 4: Earth Systems and Resources	10%–15%
Unit 5: Land and Water Use	10%–15%
Unit 6: Energy Resources and Consumption	10%–15%
Unit 7: Atmospheric Pollution	7%–10%
Unit 8: Aquatic and Terrestrial Pollution	7%–10%
Unit 9: Global Change	15%–20%

The 7 Science Practices and the exam weighting for each is as follows:

AP® Science Practices

Science Practice	Multiple-Choice Exam Weighting	Free-Response Exam Weighting
1: Concept Application	30%–38%	13%–20%
2: Visual Representations	12%–19%	6%–10%
3: Text Analysis	6%–8%	Not assessed in FRQ Section
4: Scientific Experiments	2%–4%	10%–14%
5: Data Analysis	12%–19%	6%–10%
6: Mathematical Routines	6%–9%	20%
7: Environmental Solutions	17%–23%	26%–34%

3. **Understanding the Overlap Between the Multiple-Choice and Free-Response Questions**

   Both the multiple-choice and the free-response questions are taken from topics covered in the College Board's *Course and Exam Description*. The CED contains a particularly detailed topical outline. As you study for the multiple-choice questions, you are also studying for the free-response questions. Most students fail to grasp the significance of this point. The two types of questions overlap since they both test key concepts from the same topical outline.

4. **Using Your *Crash Course* to Build a Winning Strategy**

   This *Crash Course* book is specifically designed for your success. It is based on a careful analysis of the *Course and Exam Description*'s topical outline and all the released questions. It also contains a wealth of advice based on the authors' years of experience teaching the course.

   For example, Chapter 3 contains a list of the most important key terms you need to know to be fluent in the language of environmental science. Knowledge of the vocabulary of environmental science will help you interpret the exam questions, as well as provide you with the factual information necessary to answer the questions.

   Chapters 4 through 18 provide you with a detailed discussion of each content area covered on the AP® Environmental Science exam. Use these chapters to refresh your memory on all the concepts from the course, or just focus on the specific chapters where you need the most help.

   Chapters 19 and 20 tie all of the concepts together to form a "big picture" of the relationship of humans to the environment. Review Chapters 21 through 23 right before you take the exam. These chapters include last-minute pointers on question-type strategies and present multiple-choice practice questions that will help ensure you are ready for the exam.

5. **Supplement This *Crash Course* with College Board Materials**

   Your *Crash Course* contains everything you need to know to score a 4 or a 5. However, you should also make use of the College Board's AP® Central website. It contains the *Course and Exam Description* booklet as well as free-response questions from the last 20 years.

Basic Science and Math Concepts

The AP® Environmental Science exam is an interdisciplinary science class, applying concepts you have learned in other science courses such as biology, chemistry, and geology. Many students who have taken a number of AP® courses may find the review of the basic science and math concepts relatively easy. Others with little prior exposure will need to spend more time on this review. This chapter is for those of you who need a little brush-up on basic science and math skills. Experienced AP® students should at least skim this chapter to be sure you remember some of the basics you may not have used in a while.

I. Science Review

The writers of the AP® Environmental Science exam assume that test-takers have a solid science background that includes high school-level courses in biology and chemistry. The reality is that AP® Environmental Science students are a very diverse mix. There is no "typical" AP® Environmental Science student. The goal of this *Crash Course* is to focus on specific skills and background knowledge that show up regularly on the AP® Environmental Science exam but are not part of the topic outline for the course. This will ensure that you are building your knowledge on a firm foundation.

A. Data Analysis. Expect to work with scientific data on the exam. AP® Environmental Science Practice 5 asks students to analyze and interpret quantitative data represented in tables, charts, and graphs. You should be able to describe relationships among variables and trends in the data. In addition, you must be able to relate the data to a broader environmental issue.

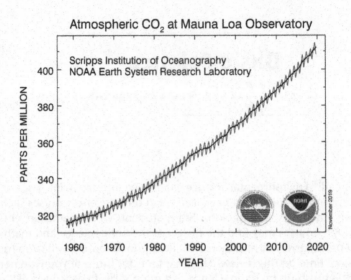

Atmospheric CO_2 at Mauna Loa Observatory

Scripps Institution of Oceanography
NOAA Earth System Research Laboratory

Data Analysis Practice

The following Data Analysis questions are based on the graph above.

1. *Which of the following statements describes the trend in CO_2 concentration from 1960-2020?*

   a) *There is a steady increase in CO_2 followed by decrease.*

   b) *CO_2 concentration and time are negatively correlated.*

   c) *CO_2 concentrations have increased over time.*

   d) *Burning fossil fuels has contributed to CO_2 concentrations.*

2. *Which of the following statements explains the yearly rise and fall of CO_2?*

   a) *Changes in rates of photosynthesis in the northern hemisphere where most terrestrial plants are located*

   b) *Increased burning of fossil fuels during winter months*

   c) *Large-scale deforestation in the southern hemisphere*

   d) *Deadzones in the Gulf of Mexico caused by algal blooms*

ANSWERS

1. c

2. a

B. World Map. Familiarize yourself with the world map. As you study things like climate and ocean currents, El Niño, human population, and distribution of the world's reserves of oil and coal, refer to a map and identify the places that you are studying. It is very common for phrases like, "this is a location of high volcanic activity," or "this is the area of the world with the greatest coal reserves," to be given and associated with a world map like the one below. The map may be labeled with the letters A–E on different regions of the globe. Familiarizing yourself with the map will make these questions much easier to answer correctly.

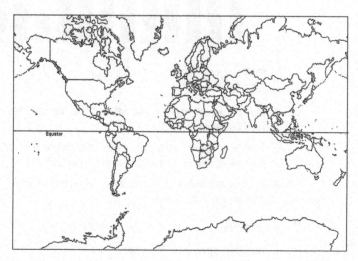

Data
Analysis
Practice

Use the climatogram below to answer the following data interpretation questions:

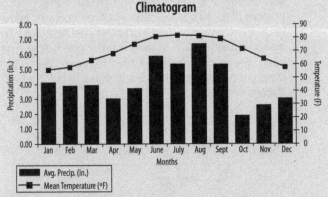

Climatogram

1. **During what month is the rainfall the highest? How much rain fell in that month?**

2. **Which of these biomes does this climatogram most resemble: desert, subtropics, boreal forest?**

3. **Is there a relationship between precipitation and temperature for this biome?**

ANSWERS:

1. **August, approx 6.5 inches.**

2. **Subtropics.**

3. **Yes, they are proportional. In the months with highest temperature, precipitation is also highest. The months with lowest precipitation also tend to have the lowest temperatures.**

C. Mathematical Routines-Calculations. You will be asked to calculate correct answers showing units.

Throughout this *Crash Course*, you will find math practice problems. These practice problems will give you a chance to see exactly what types of math questions you will encounter on the exam. For now, let's review some basic skills and tips.

1. When given a question with several numbers, don't panic!

i. First, look closely at each number and its units (e.g., miles, gallons, months). If the numbers are written into a long paragraph, pull them out and make a list to the side of each number and its unit.

ii. Next, read the question to find out exactly what is wanted in the answer. If you know the unit of the answer, you can work backwards to figure out how to set up your numbers to cancel out the units you don't need and be left with only the units you do need. This process is called dimensional analysis and it can make the most complicated question seem truly simple.

iii. Let's look at an example:

> The only time you are likely to see an adult sea turtle on the beach is during nesting season. Every two years, adult loggerhead sea turtles find mates and the females come up on to the sandy beaches to dig nests and lay their eggs. A female loggerhead will lay about 100 eggs per nest. In a nesting year, she will lay about five nests. Of those eggs, about 80% emerge as hatchlings. It is estimated that only 1 in every 1000 hatchlings survives to become an adult sea turtle.

> If loggerhead sea turtles reach sexual maturity at age 25, how many years must a female loggerhead live to produce enough adult turtles in the next generation to replace both her and her mate?

Begin by listing all of the numbers and units:

Nests every 2 years with 5 nests/nesting year

$$\frac{100\ eggs}{nest} \rightarrow 80\%\ hatchlings$$

$$\frac{1}{1000}\ hatchlings \rightarrow adulthood$$

25 years to adulthood

Then notice that you want your answer in years. Then think, "If I need two adult turtles, and only $\frac{1}{1000}$ hatchlings lives to adulthood, then I need 2000 hatchlings." Now you know that you need to find how many years it takes to produce 2000 hatchlings.

Here's one way to work it out:

$$\frac{5\ nests}{2\ years} \times \frac{80\ hatchlings}{nest} = \frac{200\ hatchlings}{year}$$

Now, how many years does it take to produce the 2000 hatchlings needed to get two adult turtles?

$$2000 \ \text{hatchlings} \times \frac{1 \ \text{year}}{200 \ \text{hatchlings}} = 10 \ \text{years}$$

(Notice that in the equation above, 1 year/200 hatchlings was flipped to cancel hatchlings and end up with years as the unit.)

So in just a few easy steps we have used dimensional analysis to keep our units straight and get the correct answer! You will find that most of the calculation-based questions on the exam will work the same way.

Be sure you isolate the numbers out of the paragraph, keep track of your units, and set up your equation to cancel out all the units you don't want to leave behind in the answer! It is easy to feel rushed and start dropping the units from your rough calculations, but this will only cause problems down the line.

2. After you complete your calculations, do a common sense check. Ask yourself, "Does this answer make sense?" For example, if your answer above had been 350 years old because of an extra stray zero, you would stop and think, "Hmmm, if loggerhead sea turtles have to live 350 years just to replace themselves, how has this species survived for millions of years?" The answer is, of course, "Impossible!" If this happens, simply go back and re-check your work to find your mistake. Did you multiply when you needed to divide (this is where writing out units and canceling out is helpful)? Did you add or drop a zero or two between steps? These are all common errors that we tend to make when we are rushing or under stress.

3. Calculating percent increase or decrease shows up in many situations. For problems like this, use the following equation:

$$\frac{\text{change in amount}}{\text{original amount}} \times 100\%$$

Subtract the smaller number from the larger number to find the change in amount (this must always be a positive number), then divide by the original amount.

Here's an example.

A scientist monitoring a population of water beetles noticed that in 2001 the population was 500 beetles. Ten years later, she counts only 300 water beetles in the population. What is the percent decrease of this population?

$$\frac{500-300}{500}=\frac{200}{500}=0.4$$

$0.4 \times 100\% = 40\%$ *decrease*

Say the population went from 500 water beetles in 2001 to 600 in ten years. What is the percent increase of the population?

$$\frac{600-500}{500}=\frac{100}{500}=0.2$$

$0.2 \times 100\% = 20\%$ *increase*

Data Analysis Practice

1. **A country in East Africa currently has a population of 22,000,000 with a growth rate of 3.5%. What will the population be in 40 years?**

2. **The Wilson family has a toilet that used 13.5 liters per flush. If they flush their toilet 10 times each day, how much water is used in liters per year, for toilet flushing per year in the Wilson house?**

ANSWERS:

1. **88,000,000**

   The rule of 70

   $$\frac{70}{2} = \text{35-year doubling time.}$$

   The population will double 2 times in 40 years.

2. **49.275 liters**

   $$\frac{13.5 \text{ liters}}{1 \text{ flush}} \times \frac{10 \text{ flush}}{1 \text{ day}} \times \frac{365 \text{ day}}{1 \text{ year}} = 49,275 \frac{\text{liters}}{\text{year}}$$

D. Experimental Design. Free-Response Question 1 is an Experimental Design Question. Remember this science practice will also be assessed in the multiple-choice section.

1. Identify a testable hypothesis or scientific question for an investigation.

   i. There are two types of variables, the independent variable (what is changed), and the dependent variable (what is observed).

   ii. A well-crafted research question will make these variables clear. The "if " part of the statement addresses the independent variable (salinity of the soil); the "then" part of the statement addresses the dependent variable (growth rate).

2. Designing the experiment. There are several things to consider when designing a good experiment.

   i. Control group. The purpose of the control group is to ensure that any changes you see in your test groups are caused by the independent variable.

      ➤ In an experiment involving the growth of bean plants in various soils, the control group would be bean plants grown in soil with no salt. If the plants grow well without salt, but poorly in the soil with salt, you can conclude that salt was the cause of the poor growth.

      ➤ If plants are grown in soils of varying salinity, but no control group is set, then a lack of growth could have been caused by the presence of salt, or perhaps there was some other condition that led to poor growth (plant disease, lack of nitrogen in the soil, too much or not enough water, etc.).

   ii. Only one variable is manipulated at a time. All groups should be identical except for the one independent variable.

      ➤ For example, the bean plants should all be the same variety, age, and approximate size. All plants should get identical amounts of water, sunlight, and soil nutrients.

The only difference should be that salt is added to some groups and not to the control.

iii. Replication. To increase the validity of the data collected from an experiment, it should be able to be replicated. So, the experiment should be done more than once to account for individual differences that may arise by chance.

➤ For example, maybe one plant just was not healthy and dies. In order to be certain its poor health was a result of its test conditions, you want to have several plants growing in that same set of conditions.

➤ There is no set number of replications that is always correct. Depending on the experiment, you may have as few as five or six replications per group or as many as hundreds of replications per group.

3. Making a graph to visualize the results. There may be an FRQ in which you are given a set of data and a blank grid and are asked to set up a graph and plot the data. To ensure you don't lose any points, be sure to consider all of the following:

i. Set up your axes so that the independent variable is always on the horizontal (x) axis, and the dependent variable is on the vertical (y) axis.

ii. Make sure you include a descriptive label, with units, for each axis.

➤ For the above experiment, the x-axis should be something like "Salinity of soil (mg/ml)."

➤ The y-axis could be "Growth of bean plants (mm)."

iii. Determine an appropriate scale for each axis so that you use at least 75% of the grid provided, and keep a consistent scale interval (count by .1, .2, .3, etc., or 10, 20, 30…).

When asked to make a graph, the dimensions of the grid given will suggest the spacing of your intervals. For example, if counting by tens is best on your y-axis, then the number of squares given for that axis will be such that if you count by tens, you will have exactly the right number of squares. Don't panic, however, if you don't end up with an exact count. There is always a range of scales that are accepted as long as they are at consistent intervals. Include all of the data and use at least 75% of the grid.

iv.   Line or bar graph?

> ➤ A line graph is used to show continuous data, such as how something changes over time. For our bean experiment, a line graph would be the best choice.

> ➤ Bar graphs are used to display discrete data. For example, if we were to count cars in a parking lot and graph them by color, each bar on the graph would represent all of the cars of a single color.

E. Feedback Loops. Every system in nature is in a nearly constant state of change. In most cases, these changes may be small adjustments that keep the system in balance. In other cases, the changes are greater and throw the system out of balance. These changes are referred to as feedback loops.

1. A *negative feedback loop* is a state of adjustment that helps keep a system stable. Think of the heating and air conditioning systems in a home. The desired temperature is set and when it gets too warm, the air conditioner kicks on to cool the house; when the desired temperature is reached, it turns off. If it gets too cold in the house, the furnace comes on to warm the house back to the desired temperature, then turns off.

2. A *positive feedback loop* happens when changing something about a system sets up a situation where that change is amplified further and further. Imagine that an increase in atmospheric temperature causes polar snow and ice to melt, which in turn causes more heat to be absorbed by the liquid, which causes more melting.

3. Positive feedback loops are often called vicious cycles because once begun, the loops are difficult to stop or reverse.

4. The terms "positive" and "negative" refer to the direction or amount of overall change in the system. Negative feedback loops contribute to overall stability (no change), whereas positive feedback loops lead to increasing change in a system.

Test Tip

Many students find this concept of feedback loops confusing because they think that "positive" must be good and "negative" must be bad. However, in most cases, the reverse is true when thinking about feedback loops.

F. Relationship of Temperature to Density and Volume. All matter in the universe is made up of atoms that combine to make molecules. The states of matter—solid, liquid, and gas—are determined by how energized the molecules of a substance become.

1. For example, if we take liquid water and place it in a very cold environment, heat is lost from the water and its molecules move less and less until they "freeze" into the crystal lattice structure we call "ice."

2. Warming this ice adds energy to its molecules. When they acquire enough energy to move around in three dimensions—sliding over and around one another—the ice melts and becomes water.

3. If we heat this water, eventually its molecules will become so energized that they spread out enough, and have such violent collisions, that they leap out of the water and become water vapor.

4. In most cases, as a substance changes from solid to liquid to gas, the molecules spread farther and farther apart as energy is added. Thus, solids tend to have the most molecules per unit of volume, liquids have fewer molecules in that same volume, and gases even fewer molecules in that volume.

5. *Density* is mass per unit of volume. Since molecules have mass, more molecules per unit of volume = greater density.

6. For most substances, solids (which exist at lower temperatures) have the highest density, followed by liquids (which exist at intermediate temperatures), and gases (which exist at higher temperatures) have the lowest density. An exception to this is water, which is most dense at 40 degrees C. This is the reason that ice floats.

7. The relationship between temperature and density applies to a substance even if the differences are not great enough to lead to a change in phase. For example, one of the key concepts driving weather patterns is that hot air tends to rise while cold air tends to sink towards the ground.

REFERENCES

Atmospheric CO_2 at Mauna Loa Observatory: *https://www.noaa.gov/news/global-carbon-dioxide-growth-in-2018-reached-4th-highest-on-record*

World Map: *https://www.freeusandworldmaps.com/html/World_Projections/WorldPrint.html*

Data for Climatogram: *http://www.weather.com/weather/wxclimatology/*

World Coal Supply Data: *https://www.eia.gov/coal/data.php*

I. Earth Systems and Resources

A. Geologic Time Scale vs. Human Scale

1. Current scientific evidence places the Earth's age at more than 4.6 billion years. When we think of the ways the Earth changes in geologic time, those changes are occurring over the course of hundreds of thousands of years, at a rate that would be undetectable in human time.

2. Human time runs on a much shorter scale. Each human generation is only about 30 years long. The major changes that humans have brought about on the Earth have all occurred within the last few thousand years. Most people think in terms of the next hundred years.

3. Many people have difficulty grasping the large spans of time between geologic events, which creates problems understanding plate tectonics, evolution, and other scientific phenomena that occur over very long spans of time.

B. Earth's Spheres

1. When studying the Earth, we often talk about its activities taking place in four spheres, or systems, which interact in complex, dynamic ways:

    i. Lithosphere

    The *lithosphere* is composed of the solid, outermost layer of the Earth. It includes the crust and upper mantle. The lithosphere is mostly rock and includes soils, silts, and sediments. It makes up the ocean floor, mountain ranges, and everything we think of as ground.

ii. Atmosphere

The gases that surround us and make life on Earth possible are known as the *atmosphere*. The air we breathe and our weather systems occur in the atmospheric layer closest to the ground called the *troposphere*. The troposphere is made up mostly of nitrogen gas (about 78 percent) and oxygen (21 percent) but also includes water vapor, carbon dioxide, and pollutants such as methane and oxides of nitrogen and sulfur. Above the troposphere, about 11 miles (17.7 km) above sea level, is the *stratosphere*, which contains the ozone layer that protects Earth from too much UV radiation. Above the stratosphere are layers called the *mesosphere, thermosphere*, and *exosphere*.

iii. Hydrosphere

All of the water on Earth is collectively called the *hydrosphere*. This includes the oceans, which cover about 70 percent of Earth's surface and contain about 97 percent of Earth's water supply, as well freshwater lakes, streams, rivers, and under-ground aquifers. Less than 1 percent of the water on Earth is available for human use, such as drinking and irrigation. Of that water, most is stored as groundwater in underground aquifers.

iv. Biosphere

➤ The *biosphere* encompasses all life on Earth. The bio-sphere spans across all three of the other spheres because there is life in soil, water, and air.

➤ *Biomass* is organic material made from living things, like wood from trees, organic content in soils, or peat harvested from bogs. Biomass materials may be burned as a source of energy.

C. Plate Tectonics

1. *Plate tectonics* is the scientific theory that explains the formation of deep ocean mountain ranges and trenches and the occurrence of volcanoes and earthquakes. The theory states that the Earth's crust is broken into several large plates that move slowly over a hot layer of the Earth called the *aesthenosphere*, the upper layer of the mantle. The movement of these plates in relation to one another causes several geological conditions, especially along plate bound-aries. There are three types of plate boundaries: the convergent boundary, the divergent boundary, and the transform boundary.

i. Convergent Boundary

> When two plates move toward one another, they eventually collide and one plate will move on top of the other, creating a subduction zone. This often results in mountain range formation and volcanic activity as seen around the so-called Ring of Fire in the North Pacific. In most cases, the overriding plates are continental crust that rides over the top of the more dense oceanic crust.

> Deep trenches, like the Mariana trench in the Pacific Ocean, also result from convergent boundaries.

> The Himalaya mountain range was formed by a convergent boundary between the Indian and Eurasian plates.

ii. Divergent Boundary

> When two plates move away from one another, a gap forms and magma from the mantle rises through the gap to form broad mountain ranges and rift valleys.

> The African Rift Valley running through Ethiopia, Uganda, Kenya, and Tanzania is caused by a divergent boundary.

> The most notable divergent boundary mountain range is the Mid-Atlantic Ridge, which runs along the ocean floor from the Arctic Ocean to near the southern tip of Africa.

iii. Transform Boundary

> When two plates slip along one another side by side, this often forms a transform fault and results in earthquakes.

> The most notable transform fault is the San Andreas Fault in southern California.

II. The Living World

A. Biotic and Abiotic

1. When discussing the interaction of living organisms with the environment, we often use the term *biotic* to mean anything that is or was living.

    i. Examples of biotic components of an ecosystem include any living plants, animals, fungi, and bacteria, as well as leaf litter, animal wastes, and remains of dead organisms.

      ii.   A biotic factor is anything that results from interactions with living things. For example, parasitism, a bacterial infection, and even competition are all biotic factors.

   2. *Abiotic* means anything that is nonliving.

      i.   Examples of abiotic components include air, water, rocks, and anything else in an environment that is not a direct product of a living thing.

      ii.   An abiotic factor is anything nonliving that affects life, such as the weather or climate, including temperature, rainfall, and sunlight.

      iii.   Other abiotic factors include soil texture and moisture, atmospheric pressure, and the presence of chemicals such as oxygen, acids, and nitrogen-containing compounds in the air, water, and soil.

B. Ecosystem

   1. An *ecosystem* is the interaction of all the living organisms in an area with their nonliving environment.

   2. Ecosystems on land are called *terrestrial ecosystems,* while those in the water are *aquatic ecosystems.*

C. Species

   1. *Species* is a group of organisms that share similar physical and behavioral traits and that can interbreed to produce fertile offspring.

      i.   Keystone Species

         ➤ In many ecosystems, one or more species are essential to the maintenance of the ecosystem. If a keystone species is removed, the whole system is at risk of total collapse.

         ➤ Top predators, like sea otters and coyotes, are common keystone species.

         ➤ Environmental engineers are a type of keystone species that physically manipulate the environment in a way that makes their ecosystem possible.

         ➤ African elephants are an example—they uproot small tree saplings in their quest to find food, thus keeping the forest from encroaching into the grassland.

   ii. Indicator Species

➤ Indicator species are especially sensitive to changes in the environment.

➤ Many frog species are indicators of water pollution because their permeable skin makes them especially sensitive to toxins.

➤ Mayflies are an indicator of high water quality in streams and rivers. They are also sensitive to pollutants, so finding mayflies and their larvae in an ecosystem indicates high water quality.

D. Habitat vs. Niche

1. Habitat and niche are related concepts that are often confused.

➤ The *habitat* of an organism is where it lives, while a species' *niche* describes its role within the ecosystem.

➤ A shark's habitat may be described as the open ocean or coral reef, while its niche would be its role as a top predator.

E. Community

1. A *community* is the assemblage of all organisms living and interacting in a particular area.

➤ A community includes all of the different species of plants, animals, fungi, bacteria, and protists.

➤ Add the nonliving environment to a community and you have an ecosystem.

F. Trophic Levels

1. *Trophic levels* are also known as feeding levels.

   i. The first trophic level in any ecosystem is the producers—autotrophs that make their own food. In most cases, autotrophs are photosynthetic plants or chemosynthetic bacteria.

   ii. The next trophic level is the primary or first-degree consumers—herbivores (plant eaters) that feed on producers.

   iii. Secondary or second-degree consumers are the omnivores and carnivores that feed on primary consumers. This may continue to tertiary (third-degree) consumers feeding on secondary consumers and quaternary (fourth-degree) consumers feeding on tertiary consumers.

    iv. Most ecosystems have no more than four or five trophic levels due to the fact that only 10 percent of the energy available at one trophic level is transferred to the next highest level.

G. Biomes

    1. A *biome* is a large region of the Earth characterized by a distinct set of climate conditions.

    2. Terrestrial biomes are determined primarily by temperature and precipitation.

    3. Major biomes of the Earth include forests, grasslands, deserts, wetlands, and estuaries.

H. Reservoirs

    1. A reservoir is a zone of storage or containment.

    2. When you study the biogeochemical cycles, like the carbon, nitrogen, water, phosphorus, and sulfur cycles, an understanding of reservoirs is important.

    3. For example, the greatest reservoir for carbon is the calcium carbonate in limestone deposits, but fossil fuels are also a significant carbon reservoir.

    4. A reservoir can also be a large human-made lake of water that accumulates behind a large dam. These reservoirs are often used to supply water to nearby urban areas.

 III. **Population**

A. Anthropogenic

    1. The term *anthropogenic* refers to any change in the Earth's systems caused by human activities.

    i. Most pollution is anthropogenic.

    ii. Cultural eutrophication is an anthropogenic increase of nutrients in an aquatic ecosystem.

    iii. Deforestation for agriculture and urbanization are also examples of anthropogenic changes.

2. As human populations continue to increase, we are having greater and greater anthropogenic effects on the Earth's systems.

B. Per Capita

1. *Per capita* means "per person."

2. Per capita income for a nation is a measure of how much, on average, each person in that nation earns.

IV. Land and Water Use

A. Environmental Impacts, Consequences, or Implications

1. An *environmental impact* is some change in the environment that occurs as a result of some type of activity.

➤ Every living organism interacts with the environment, resulting in environmental impacts.

➤ Possible environmental impacts of humans burning fossil fuels include habitat destruction and water acidification from the mining process, and air pollution from the burning process.

Often the terms consequences *and* implications *are used interchangeably in AP® Environmental Science exam questions. This is because essentially they mean the same thing: how are our human activities affecting the environment?*

B. Arable Land

1. *Arable land* is any land suitable for agriculture.

➤ One ever-increasing problem will be our limited amount of arable land.

➤ As populations grow, we may not have enough arable land to grow the crops needed to feed Earth's people.

C. Irrigation

1. Most crops grown for food need a steady supply of water. To increase the reliability of harvests and to produce more food per acre, we often supplement the natural rainwater that falls with

additional water taken from groundwater, lakes, or rivers. This process of watering crops is called *irrigation*.

i. Salinization

➤ Salinization is a common problem associated with irrigation. While rainwater contains few impurities, water from land-based sources will always have a small amount of dissolved salts resulting from contact with sediments and soils. When we irrigate crops, these salts build up in the soil surrounding the plants and gradually make it salty. This is a particular problem in arid areas with little rainfall because rainfall works to wash away these accumulated salts.

ii. Desalinization (or Desalination)

➤ One solution is desalinization, which dissolves the accumulated salts in water by flooding the affected soil and draining the salty water away.

iii. Water Diversion

➤ One of the original ways humans irrigated crops was to divert water from a lake or stream using a series of canals that directed the flow of water toward crops. Water diversion projects are still in use today all over the world, only now the scale is much larger.

➤ The shrinking Aral Sea in Russia is a famous example of the problems that can result from large-scale water diversion projects.

D. Extraction

1. The term *extraction* refers to mining, drilling, and removing trees from a forest. Any time we remove a resource from the Earth, it is referred to as extraction.

V. Energy Resources and Consumption

A. Laws of Thermodynamics

1. The laws of thermodynamics describe how energy behaves in Earth's systems. Make sure you understand both the first and second laws and how they apply to energy.

    i.   First Law of Thermodynamics

➤ The law of conservation of energy states that energy can neither be created nor destroyed. As a result, all the energy in the universe is held constant.

➤ Energy may be transferred from one object to another in many forms.

    ii.   Second Law of Thermodynamics

➤ The law of entropy states that energy always moves from a more concentrated state to a less concentrated state. This is why energy moves from hot objects toward cool objects (and never in the opposite direction).

➤ As a result, all natural processes involving energy transfer are linear and cannot be reversed.

➤ The sun is a one-way source of energy for the Earth. Energy moves through ecosystems from the producers at the base to the top predators at the apex.

➤ The second law of thermodynamics also explains why 90 percent of energy is lost at each step in the food chain, and why your car needs to be refilled with gas to keep going. At each conversion from one form to another, some of the energy is degraded, or lost to heat and/or friction.

    iii. Efficiency

➤ *Efficiency* is a measurement of how much energy used in a reaction is converted to useful activity and how much is wasted. A highly efficient reaction wastes very little energy.

➤ The second law of thermodynamics states that it is not possible to have a 100 percent efficient energy conversion.

B. Potential and Kinetic Energy

1. *Potential energy* is stored energy that can be unleashed to do useful work.

2. *Kinetic energy* is energy in motion.

➤ A lump of coal represents potential energy. The energy stored in the carbon and hydrogen bonds in that lump of coal can be converted to kinetic energy by burning the coal.

VI. Pollution

A. *Pollution* is anything added to the environment that may cause harm.

1. Air Pollution

   i.   *Air pollution* is any gases, particulate matter (smoke and dust), and biological entities (pollen) that are released into the atmosphere that may be harmful to life.

   ➤ *Primary pollutants*, such as NO_x, SO_x, particulate matter, and CO_2, are released directly from a contaminating source (i.e., a coal-burning power plant or factory).

   ➤ *Secondary pollutants*, including ozone, photochemical smog, and the nitric and sulfuric acids that become acid deposition, are formed by chemical reactions in the atmosphere involving primary pollutants.

2. Noise and Light Pollution

   i.   Many people do not realize that noise and light can be considered forms of pollution. Excess levels of either noise and/or light may cause disruption or harm to ecosystems, as well as to human health and well-being.

3. Water Pollution

   i.   Water pollution includes any pollutant added to water that results in harmful effects to aquatic ecosystems or water purity (for drinking and irrigation).

   ➤ *Point source pollutants* are those that can easily be identified. Point source water pollutants include effluent overflows from a wastewater treatment plant or a drainage pipe from a factory.

   ➤ *Nonpoint source pollutants* usually pose a greater problem because they are dispersed, difficult to locate, and as a result are more challenging to prevent.

   — *Runoff* is the most common example of nonpoint source water pollution. Types of runoff include nutrient-rich agricultural runoff of fertilizers and/or livestock wastes or urban runoffs containing oils and fluids used on and in cars and trucks.

— Methods to reduce runoff and increase infiltration include installing permanent pavers, planting trees, installing rain barrels, and building up, not out.

— Methods to increase water infiltration include replacing traditional pavement with permeable pavement, planting trees, and increased use of public transportation.

4. Solid Waste

   i. Most items thrown away are considered solid waste. Categories include paper, metals, glass, plastics, and organic materials.

   ii. Much of the solid waste we produce is *recyclable*, meaning it can be collected, melted down, and remade into something new.

5. Toxicity

   i. *Toxicity* is a measure of the harm a substance has on living organisms.

   ii. Toxicity of a substance is determined by a dose/response relationship. This means that, as an organism is exposed to greater amounts or concentrations of a toxin, its harmful effect increases in a predictable manner.

   iii. Chemicals can have a wide range of effects on our health. Depending on how the chemical will be used, many kinds of toxicity tests may be administered. One measure is LD50 (or LC50 in aquatic systems), the lethal dose of a toxin that will result in death to 50 percent of a test population.

6. Hazardous Waste

   i. Hazardous waste includes substances that are highly flammable, poisonous, or corrosive.

       ➤ Radioactive nuclear wastes are a class of hazardous wastes.

       ➤ *Carcinogens* are a class of cancer-causing hazardous wastes. Others may cause birth defects (*teratogens*) or genetic mutations (*mutagens*).

       ➤ All hazardous wastes require special disposal because environmental contamination by these substances leads to dramatic and often catastrophic environmental consequences.

7. Bioremediation

    i. *Bioremediation* refers to safely removing certain pollutants or hazardous chemicals from the environment and stored in the tissues of living organisms.

VII. Global Change

A. Biodiversity

1. Biodiversity is a measurement of the variety of living organisms in a system. Not only is it a count of the number of different species (species richness), it is also a measurement of the relative abundance of each species (species evenness).

2. Types of biodiversity include species diversity, genetic diversity, and ecological diversity.

B. Ozone (O_3)

1. Ozone is either a blessing or a curse depending on where in the atmosphere it is located.

    i. *Stratospheric ozone* (good ozone) is essential to life as we know it. It acts as a sunscreen to block excess UV radiation from penetrating into the troposphere, where it can lead to reduced rates of photosynthesis in plants, as well as elevated risks of skin cancer in people and other animals.

    ii. *Tropospheric ozone* (bad ozone) is a harmful pollutant, a by-product of the combustion of fossil fuels and biomass. It is corrosive, and it may cause respiratory irritation and degrade materials like plastics, metals, and fabrics.

    ➤ Ozone is also a component of photochemical smog.

C. Global Climate Change

1. Global climate change is the collection of changes in global temperatures, climatic patterns, and atmospheric events that has been attributed to human activities over the last few hundred years—especially since the industrial revolution. Most of these changes are thought to result from the increase in atmospheric greenhouse gases from burning fossil fuels.

i. *Greenhouse gases*, like carbon dioxide, methane, water vapor, nitrous oxide, and chlorofluorocarbons, are gases that trap heat in our atmosphere and cause a warming effect called *the greenhouse effect*.

ii. The greenhouse effect is essential to life as we know it. Without greenhouse gases, Earth would be a frozen planet.

iii. An increase in the concentration of greenhouse gases may cause global temperatures to increase faster than our natural systems can adapt. This threatens the existence of our ecosystems and ultimately our ability to survive on Earth.

PART II

LIFE ON
EARTH

Geology Concepts

Test Tip

Be aware of the interactions between Earth systems and geology concepts. Most geologic processes take a long time to occur. Even an event like an earthquake or erupting volcano that appears to happen suddenly is usually the result of energy that has been building up within the Earth's crust for hundreds or even thousands of years.

I. Reasons for Seasonal and Climate Variation

A. Earth's Tilt

   1. The Earth is tilted at approximately 23.5 degrees on its axis.

B. Earth's Rotation

   1. Every year (365.25 days), the Earth makes one full rotation around the sun.

C. Seasons on Earth

   1. Look at the following diagram. During one part of the Earth's orbit around the sun, the tilt causes the southern hemisphere to receive more direct sunlight. During this time, it is summer in the southern hemisphere and winter in the northern hemisphere. As Earth continues to travel around the sun, it eventually reaches a position where the northern hemisphere receives more direct sunlight (imagine the diagram below where the sun and the Earth switch positions—the sun is on the left and the Earth is on the right but still tilted in the same direction). During this time of the year, it is summer in the northern hemisphere and winter in the southern hemisphere.

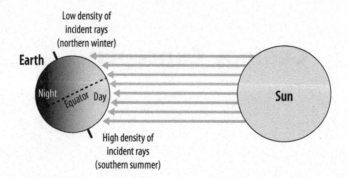

Low density of incident rays (northern winter)

Earth

Night

Equator Day

Sun

High density of incident rays (southern summer)

D. Solar Intensity and Latitude

1. Using this same concept, you can see that the equator is always receiving the most direct sunlight and the poles are always getting sunlight at a small angle.

2. Because the rays at small angles tend to spread out over larger areas of the planet, and because the more direct rays tend to be absorbed more directly, the equator is always warmer than the poles.

3. This amount of sunlight shining on various points of the Earth is referred to as solar insolation (incoming solar radiation). It is the Earth's main source of energy and is dependent on season and latitude.

II. Plate Tectonics

A. The Earth's crust is broken up into large segments called *tectonic plates.*

1. These plates are in constant (but very slow) motion and interact with one another in a few predictable ways at their boundaries.

B. The Earth's plates come together in three major types of boundaries: divergent, convergent, and transform.

1. *Divergent plate boundaries,* also called spreading centers, are formed when two plates are moving away from one another. This spreading center causes magma from the Earth's interior to rise and cool, making new crust in the form of undersea mountains.

      i.   Earth's most notable divergent boundary is found at the mid-Atlantic ridge running along the seafloor of the Atlantic Ocean between the Americas, Africa, and Europe.

      ii.   In general, divergent plate boundaries are associated with seafloor spreading, deep ocean mountain ranges, Africa's rift lakes, and volcanic activity.

2. _Convergent plate boundaries_ are formed when two plates meet and grind together. The plate that is more dense is usually pushed beneath the less dense plate, forming a subduction zone.

      i.   The so-called Ring of Fire is a famous pattern of volcanic and earthquake activity along the continental borders of the Pacific Ocean. This area where the Pacific Plate is subducting beneath surrounding plates accounts for almost 75 percent of Earth's volcanoes.

      ii.   The Marianas Trench (Earth's deepest crevice) is part of the Ring of Fire and is an example of a convergent boundary between two oceanic crust plates.

      iii.   If both plates are made primarily of less dense continental crust, neither will subduct. Instead, the plates collide and buckle upward into a massive mountain range.

      iv.   The most magnificent example of this is the formation of the Himalaya Mountain range as the Indian plate collides with the Eurasian plate.

      v.   In general, convergent plate boundaries are most often associated with deep ocean trenches, volcanoes, and earthquake activity.

3. _Transform plate boundaries_ are formed when two plates, or fractures in a single plate, slide horizontally along one another.

      i.   Often called a transverse fault, the San Andreas Fault, which runs along southern California and northern Mexico, is one of the most notable transverse boundaries.

      ii.   Transform boundaries are most often associated with earthquakes.

C. Intra-plate hot spots are areas of the Earth where there is volcanic activity that cannot be explained by plate boundaries. They are thought to be formed by mantle plumes of rising hot magma beneath one of the plates, which causes softening of the plate and eventually allows magma to break through in the form of a volcano.

1. The Hawaiian Islands are the most notable examples of intraplate hot spots. Repeated volcanic eruptions cause a buildup of lava rock, which eventually rises above sea level and forms exposed islands.

The only topic from this chapter that regularly appears on the AP® Environmental Science exam is that the seasons are caused by the tilt of the Earth on its axis as it rotates around the sun. Be sure to know this. Plate tectonics shows up occasionally, so it's important to know that volcanoes and earthquakes are mostly caused by interactions at plate boundaries. Be prepared to identify the location of significant geologic features on a map.

Interactions Between Earth's Systems: Air and Water

I. Composition of Earth's Atmosphere

A. Layers of the Atmosphere

  1. The *troposphere* is the layer where we live; most of the atmosphere's oxygen is found here.

    i. The troposphere extends upward 5 to 9 miles from the Earth's surface at sea level.

    ii. Most weather occurs in the troposphere.

    iii. The troposphere is the most dense layer of the atmosphere (more molecules per unit of volume).

  2. The *stratosphere* is directly above the troposphere.

    i. The stratosphere extends from 9 to 31 miles above the Earth's surface at sea level.

    ii. Stratospheric ozone is vital to trapping and scattering ultraviolet (UV) radiation from the sun (see Chapter 14 for a full discussion of ozone).

    iii. Compared to the troposphere, the stratospheric air is drier and less dense.

    iv. The troposphere and stratosphere together make up the lower atmosphere. Ninety-nine percent of all atmospheric air is found here.

  3. The *mesosphere* lies directly above the stratosphere from 31 to 53 miles above the Earth's surface at sea level.

  4. The *thermosphere* is directly above the mesosphere, extending from 53 to 372 miles above the Earth's surface.

5. The *exosphere* is the outermost layer of Earth's atmosphere and acts as the transition zone between Earth's atmosphere and outer space.

   i.   Of all the layers, the exosphere has the lowest atmospheric pressure.

B. Composition of the Atmosphere

1. The atmosphere is made up of gases. The table below shows the percentage of each gas in the atmosphere.

Atmospheric Gas	Percentage of the Earth's Atmosphere
Nitrogen (N_2)	78
Oxygen (O_2)	21
Water vapor (H_2O)	0–7 (varies by climate)
Carbon dioxide (CO_2)	0.01–0.1 (varies by location)
Ozone (O_3)	0–0.01 (varies by location)

C. Greenhouse Effect

1. The temperature of Earth's atmosphere is maintained by the absorption of incoming infrared radiation (heat) by atmospheric gases such as CO_2 and CH_4. (See Chapter 17 for a full discussion of the greenhouse effect.)

D. Thermal Inversion

1. A *thermal inversion* occurs when the air temperature at the Earth's surface is cooler than the air at higher altitudes.

   i.   In most cases, the air of the lower atmosphere is warmest near the Earth and becomes cooler as it gets farther from Earth's surface. The reason is that Earth absorbs infrared radiation (heat) from the sun and radiates it back into the surrounding air.

   ii. Inversions may also lead to smog when dust and particulate pollutants collect in the cooler, still air.

       ➤ A persistent atmospheric inversion in an area with smoke from forest fires can cause the smoke to be

> trapped, leading to air quality problems for several days.

> ➤ Atmospheric inversions are commonly found in large cities that are surrounded by hills and mountains, such as Los Angeles; Mumbai, India; Mexico City, Mexico; and Vancouver, British Columbia.

II. What Causes Weather?

A. Definition

1. *Weather* is the day-to-day temperature, atmospheric pressure, and precipitation of a given location.

B. Temperature and Density

1. When thinking about weather, remember that temperature and density are closely related. (See Chapter 2 for a more detailed discussion of temperature and density.)

    i. Cold air is more dense (more molecules per unit of area). Because of its density, cold air tends to sink through warmer, less dense air. Conversely, warmer air is less dense and rises through cooler air.

    ii. As warm air rises through cooler air, it loses heat and cools. Because cooler air has a lower capacity for holding water vapor, any vapor in the warmer air condenses into water as it rises. Thus, warm air rising tends to return water to the ground in the form of precipitation (rain, sleet, snow, or hail).

    iii. As cool air sinks through warm air, it gains heat and warms up. Warm air has a greater capacity for holding water than cool air. As cool air descends toward the ground, it tends to take moisture up and away from the ground as the air warms.

C. Global Wind Patterns

1. Global wind patterns primarily result from the most intense solar radiation arriving at the equator, resulting in density differences and the Coriolis effect (how a moving object such as a hurricane seems to veer toward the right in the northern hemisphere and is deflected toward the left in the southern hemisphere).

III. Earth's Geology and Climate

A. Climate

   1. *Climate* is the long-term pattern of weather conditions of a particular area.

B. How Earth's Atmospheric Convection Cells Relate to the Biomes

   1. In the following figure, look at the equator. As the air is warmed by the sun's radiation, it rises and cools. As the warm, moist air rises and cools, the water vapor it contains condenses and falls as rain. This is why tropical rainforests are common along the equator.

   2. At 30° latitude (north and south of the equator), notice that the now cool air falls toward the Earth. As that air falls and warms, it has a greater capacity for holding water, and it absorbs water from the Earth's surface, carrying that moisture farther from the equator to 60° latitude. This is why it is common to find deserts at 30° latitude.

   3. At 60° latitude, the warm, moist air once again rises and cools, dropping water toward Earth's surface and causing temperate rainforests to be common at this location.

   4. The final rotation of the cells results in cool, dry air descending over the poles, warming and pulling water from the land and leading to polar deserts.

   5. Combine the idea of the atmospheric convection cells with the fact that the equator is warmer due to more direct infrared (IR) radiation from the sun and with the fact that the poles are colder due to indirect IR radiation, and you can account for the full array of biomes found on Earth.

Test Tip *Although you need not know the names of all of the wind patterns, understand the big concept: Most of the Earth's atmospheric processes are driven by input of energy from the sun.*

The Convection Cell Model of Atmospheric Circulation

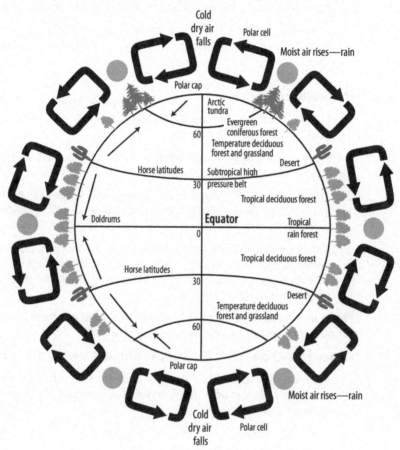

C. Geologic and Geographic Factors

  1. Weather and climate are affected not only by the sun's energy but by geologic and geographic factors, such as mountains and ocean temperature.

  2. A rain shadow is a dry area on one side of a mountain. Water vapor in the air condenses and falls as precipitation as it rises in altitude. But, the height of the mountain prevents the rain from moving to its dry side.

D. Ocean Interactions

  1. Convection currents in ocean water work in nearly the same manner as atmospheric convection cells. In general, cool dense water

from the poles sinks and moves toward the equator, while warmer, less dense water near the equator rises and floats toward the poles.

2. Because water has an even greater capacity for holding heat than air, these ocean currents carry much of the Earth's heat from place to place.

E. El Niño Southern Oscillation

1. The El Niño Southern Oscillation (ENSO) is a climate shift found in the tropical equatorial Pacific Ocean.

2. El Niño occurs when the trade winds blowing east to west across the equatorial Pacific Ocean slacken or reverse.

  i. This causes surface waters to be warmer and suppresses nutrient rich upwellings off the northwest coast of South America (near Peru).

  ii. If the upwellings are halted for more than 12 months, populations of plankton, seabirds, fish, and other critters find it difficult to find enough nutrients to survive. As a result, their population numbers decline.

3. ENSO events occur every three to seven years and last from a few months (normal) to as long as 4 years (extreme)!

4. A strong ENSO can affect weather patterns over most of the Earth.

  i. Warmer, dry weather may be found across much of the northern U.S., Canada, Brazil, Indonesia, Australia, India, and southeast Africa, leading to:

   ➤ less snow, causing decreased snowmelt in the spring to replenish snow-fed rivers

   ➤ severe droughts

   ➤ increased wildfires

   ➤ increased transmission of diseases due to stagnant warm water and lack of winter kill-off of insect vectors (malaria, dengue fever) and deterioration of fresh water in times of drought (concentration of pollutants, cholera, and other diseases)

  ii. Significantly higher rainfall may occur across much of the southern U.S., Cuba, northern Peru, Ecuador, Bolivia, southeast Argentina, and equatorial east Africa, leading to:

   ➤ suppression of hurricanes in the Caribbean and Atlantic Ocean

➤ flooding which is often followed by landslides

➤ increased transmission of diseases (amoebic dysentery, cholera, giardia) due to contamination of water sources because of flooding and standing water (mosquito breeding).

Test Tip

Past AP® exams have focused on the environmental changes and effects that result from El Niño or its colder counterpart, La Niña.

IV. Global Water

A. Agricultural Use of Water

1. Agricultural use is any use of water to grow crops or to maintain livestock. Almost 70 percent of the world's freshwater use is for irrigating crops.

2. Irrigation is most effective during the cooler parts of the day (early mornings and evenings). This reduces the amount of water lost to evaporation.

3. There are many methods of irrigation, including passive, spray, and drip irrigation.

   i. *Furrow irrigation* involves digging trenches or furrows between rows of crops and allowing water to flow to the crops. This system is inexpensive; drawbacks include loss of water to evaporation and runoff.

   ii. *Spray irrigation* involves using sprinklers or sprayers to water plants. This method is relatively inexpensive to install, is mobile, and can water many plants with relatively few sprinklers. The major drawback is that much water is lost to evaporation, especially in hot dry air.

   iii. *Drip irrigation* is often accomplished by using soaker hoses to deliver a slow, steady supply of water directly to the roots of plants. This is especially effective at reducing evaporative water loss, but it is expensive and labor-intensive to install.

   iv. Flood irrigation is a low-tech method of irrigation in which an entire field is covered with water. Runoff and evaporation are also problems associated with flood irrigation, in addition to water-logging.

4. Water-logging occurs when excessive irrigation raises the water table of groundwater and inhibits plants' ability to absorb oxygen.

5. Salinization results when crops are irrigated and salts remain in the soil after irrigation water has evaporated. The salt can build to levels that are toxic to the plant.

6. Aquaculture (raising fish, mollusks, and crustaceans for food, often in large land-based tanks) is another agricultural use of water.

B. Domestic Water Use

1. Water can be conserved in the home by using drought-tolerant plants when landscaping; using washing machines and dishwashers only when the machines are fully loaded; reducing the frequency and length of showers; turning off the faucet when brushing teeth; and installing low-flow toilets, faucets, and showerheads.

C. Legislation and Conservation

1. The Clean Water Act

   i. The objective of the Clean Water Act is to restore and maintain the chemical, physical, and biological integrity of the nation's waters. This is accomplished in three ways:

   ➤ Preventing point and nonpoint pollution sources

   ➤ Helping publicly-owned water treatment plants to improve wastewater treatment

   ➤ Working to maintain the integrity of wetlands

2. The Safe Drinking Water Act

   i. The Safe Drinking Water Act protects the quality of drinking water with two major provisions that:

   ➤ protect current and potential sources of drinking water from contamination and/or pollution, including surface water and groundwater.

   ➤ set and enforce minimum standards of purity for tap water provided by wells and municipal water companies.

Frequently, the free-response questions on the AP® Environmental Science exam will involve a specific piece of legislation. If you are asked to discuss legislation relating to freshwater pollution, or domestic, industrial, or agricultural water-use issues, the Clean Water Act and the Safe Drinking Water Act are your best bets. One of these two could be applied to almost any situation.

REFERENCES

Three Cell Model of Atmospheric Circulation. *http://www.life.umd.edu/classroom/biol106h/L34/L34_climate.html*

Clean Water Act definition from the Environmental Protection Agency. *http://www.epa.gov/agriculture/lcwa.html*

Interactions Between Earth's Systems: Soils and Cycles

I. Properties of Soils

A. Soil Formation

    1. Weathering

        i.   Soils are formed when parent material (various sediments) is weathered, transported, and deposited.

        ii.   *Weathering* is the gradual breaking down of rock into smaller and smaller particles.

        iii. *Parent physical weathering.* Parent materials can be broken down by physical, chemical, and biological processes.

    2. Erosion

        i.   *Erosion* is the movement of soil or rock particles from one place to another. Soils can be eroded by wind or water.

        ii.   Soils provide valuable ecosystem services of filtering and cleaning water that moves through them.

B. Physical Properties of Soils

    1. *Texture* is determined by the particle sizes represented in the mineral component of a soil sample. Particles are classified by size as sand, silt, or clay.

        i.   A soil triangle is a diagram that classifies soil based on the percentages of sand, silt, and clay.

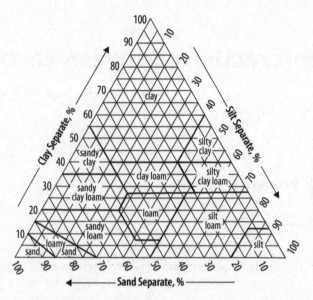

2. *Porosity* is determined by the amount of air space between the particles of soil. Soils with high porosity have larger and/or more air spaces. This allows for development of plant roots and faster infiltration and drainage of water.

3. The moisture/water-holding capacity is the ability of soils to hold water in place.

   i. Soils with a low water-holding capacity dry out very fast after water is applied.

4. *Permeability*, or the percolation rate, is essentially the opposite of water-holding capacity. The higher the water-holding capacity, the lower the permeability (water drains through slowly).

   i. Soils with a low water-holding capacity have a high permeability.

5. *Compaction* is essentially the opposite of porosity. Soils with fewer and/or smaller air spaces are more compact.

Summary of the Physical Properties of Soil

	Sand	Silt	Clay
Particle size (mm)	2.00–0.05	0.05–0.002	Smaller than 0.002
Porosity	High	Medium	Low
Permeability	High	Medium	Low/ impermeable
Water-holding capacity	Low	Medium	High/ impermeable
Compaction	Low	Medium	High

C. Chemical Properties of Soils

1. *Organic content* includes leaves, animal wastes, and any materials derived from living (or dead) organisms. Soils with high organic content tend to be more fertile because the decay of organic material returns nutrients to the soil.

2. *Fertility* is a measure of essential nutrients (nitrogen, phosphorus, and potassium) found in a soil sample.

   i.   Nitrogen (N), phosphorus (P), and potassium (K) are the most commonly measured and supplemented soil nutrients, but there are also several micronutrients that are needed in small quantities.

   ii. Organic fertilizers add decayed organic material like composted plants or animal wastes.

   ➤ Decayed organic material increases fertility gradually as the materials decompose.

   ➤ Organic fertilizers also supply the full range of micronutrients and aid in the maintenance of good soil texture.

   iii. Inorganic fertilizers are useful because farmers can target specific soil needs and add only the necessary chemicals.

   ➤ Inorganic fertilizers release nutrients immediately, which can also lead to depletion of micronutrients and soil compaction.

3. *pH* is a measure of the acidity or alkalinity of the soil.

   i. The normal pH range for growing plants is from 6 (slightly acidic) to 8 (slightly alkaline), although some plants have an optimal pH in either the acidic or alkaline range.

Test Tip

An exam question may provide data from different soil samples, and students are asked to draw conclusions about impacts and management practices based on the data.

D. Arability

   1. *Arability* is the capacity of land to be used for growing crops. To be arable, land needs to have soil that is workable (not compacted) and fertile.

Test Tip

Soil questions frequently appear on the exam. Topics most commonly addressed are: soil horizons (diagram and characteristics of each horizon); poor soil in tropical rainforests for growing crops (and why); types of agricultural practices that lead to soil degradation; and agricultural practices that conserve soil quality.

II. Soil Profiles

A. Soil Horizons

   1. Soil layers are divided into *horizons* based on physical and chemical characteristics, as shown in the following diagram.

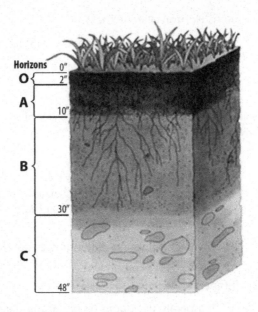

i. O Horizon (Humus)

   ➤ The topmost layer of soil is made up of partially decomposed leaf litter and other plant materials.

   ➤ O Horizon has a very high organic content.

ii. A Horizon (Topsoil)

   ➤ The A Horizon is the most fertile layer of soil.

   ➤ The A Horizon is high in organic content from the decomposition of materials in the O layer.

   ➤ Most biological activity is found in the A Horizon.

iii. B Horizon (Subsoil)

   ➤ The B Horizon layer accumulates iron, clay, aluminum, and other water-soluble compounds that are leached from the O and A horizons as water percolates down through the soil.

   ➤ Deeper plant roots also penetrate into the B Horizon.

iv. C Horizon (Parent Rock)

> ➤ The C Horizon layer contains large unweathered rocks.

> ➤ As rock particles break down, they become integrated into the B Horizon.

B. Soil Profiles by Biome (discussed in order of soil fertility, from high to low)

1. Grasslands

i. Grasslands tend to have the most well-developed soils that are rich in organic content with thick O and A horizons. For this reason, they are the biome that has been most converted to agricultural land globally.

ii. The decomposition of the grasses during the winter months replenishes the organic content of the soil annually.

2. Temperate Deciduous Forests

i. Temperate deciduous forests accumulate a thick layer of leaf litter in the fall when leaves drop. This leaf litter and dead plant matter decompose, cycling nutrients back into the soil.

3. Tropical Rainforests

i. Due to the high temperatures and rainfall, decomposition in tropical rainforests is very rapid.

ii. Most nutrients from organic decomposition are used immediately to support new plant growth, nutrients exist above the soil in the rainforest's living organisms, making rainforest soils nutrient poor.

III. **Soil—Sustainable Practices in Soil Conservation**

A. Sustainable Agriculture

1. Effects of erosion

i. Erosion leads to loss of topsoil and reduced arability (desertification).

2. Preventing Erosion

i. *No-till agriculture.* Instead of stirring up soil and turning soil over before planting, seeds are inserted into small holes or slits in the ground to reduce disturbance of the soil's O horizon.

ii. *Contour or terrace planting.* Contour (terrace) planting is used when farming on sloped land. Cutting steps or planting on a slope of land reduces the rate of water runoff.

iii. *Agroforestry.* Trees are planted in alternate rows with crop plants, or around the borders of plantings, in order to act as windbreaks and reduce water and wind erosion.

iv. *Soil coverage.* When harvesting crops like corn or wheat, the cut plant material is left to decay on the field. In seasons when a field is not planted, plant-cover crops, like native grasses or nitrogen-fixing legumes, are planted to hold soil in place. Laying down a layer of mulch can also be helpful.

v. *Special irrigation methods.* Methods such as drip irrigation reduce pooling or runoff.

vi. *Perennial crops.* These crops grow year-round and can be harvested multiple times. This reduces the need for plowing and herbicide application.

vii. *Crop rotation* is a practice in which different types of crops are planted in successive seasons, or rotated.

B. Desertification

1. *Desertification* describes the process of land becoming less suitable for cultivation—more desertlike. Areas with low rainfall (such as grasslands) are especially vulnerable.

2. Intensive farming techniques requiring high amounts of irrigation and inorganic (commercially produced) fertilizers can cause desertification. Overgrazing on semi-arid grasslands and rangelands is another common cause.

3. Procedures to reduce erosion are also effective at slowing or preventing desertification.

IV. Biogeochemical Cycles

A. Law of the Conservation of Matter

1. The law of the conservation of matter states that matter may not be created or destroyed. Matter changes forms and may chemically recombine, but the amount of matter in the universe remains constant.

2. All forms of matter cycle through the environment and take on a variety of forms.

B. Nitrogen Cycle. The nitrogen cycle traces the movement of atoms and molecules of nitrogen between various sources and sinks. Nitrogen compounds tend to cycle quickly, between reservoirs. Nitrogen is the most abundant gas in the atmosphere. Nitrogen is often a limiting factor for life in ecosystems because all living organisms need nitrogen to build essential molecules such as nucleic acids (DNA and RNA) and proteins (made up of amino acids). The problem is that atmospheric nitrogen (N_2 gas) is not very chemically active, so it is not a form that can be used by organisms. As a result, a series of chemical transformations is necessary to convert N_2 gas into more usable forms, and then ultimately back into N_2 gas again to continue the cycle.

1. *Nitrogen fixation.* Soil bacteria and *rhizobium* bacteria living in root nodules of leguminous plants (beans, peas, peanuts, and soy) convert gaseous nitrogen (N_2) from the atmosphere into ammonia (NH_3).

2. *Assimilation.* Plant roots take up ammonium ions (NH_4^+) and nitrate ions (NO_3^-) for use in making essential molecules.

C. Carbon Cycle. The carbon cycle shows the movement of carbon atoms and molecules between various sources and sinks. Like nitrogen, most carbon is found in the gaseous phase in the atmosphere.

1. Carbon cycles proceed slowly, over millions of years through chemical reactions and tectonic activity.

2. The rapid carbon cycle is an annual cycle involving the processes of photosynthesis and respiration.

    i.   Photosynthesis. Plants absorb carbon dioxide (CO_2) from the air and use it in the process of photosynthesis, eventually releasing oxygen as a by-product.

    ii. Respiration. All aerobic organisms (animals, plants, bacteria, and fungi) need oxygen for the process of cellular respiration. CO_2 is released as a by-product of the respiration process.

3. *Fossil fuels.* Plant and animal decomposition over millions of years store carbon as fossil fuels. When burned, fossil fuels release that carbon into the atmosphere as CO_2, a greenhouse gas.

4. Carbon sinks include forests, oceans, and carbonate deposits. Carbon sinks remove carbon from the atmosphere.

D. Water Cycle (Hydrologic Cycle). Ninety-seven percent of Earth's water is found in the oceans. This is the major reservoir for water, but water is constantly changing form and moving from one area of the Earth to another. The water cycle is primarily driven by the process of evaporation.

1. *Evaporation* is when liquid water changes into water vapor (gas). Heat from the sun is the major source of energy for evaporation. About 90 percent of all the water vapor in the atmosphere comes from the evaporation from surface water (oceans and freshwater) and land.

2. *Transpiration* occurs when water enters the atmosphere from plant leaves. This accounts for the remaining 10 percent of atmospheric water vapor.

3. *Condensation* occurs when atmospheric water vapor changes to liquid form. This process is how clouds and fog form. Both are made up of tiny droplets of water suspended in the air. Condensation also leads to precipitation.

4. *Precipitation* is any type of falling H_2O. This includes rain (liquid water), sleet (freezing rain), snow (partially frozen water), and hail (balls of ice).

5. Runoff (ocean storage). Much of the water that falls to the Earth lands in the oceans, lakes, rivers, or streams. The water that falls to land will either be absorbed by the land (see the discussion of infiltration in this chapter) or be returned to a body of water as runoff. *Runoff* is when falling water flows across the land (usually pulled by gravity) and eventually makes it back to an ocean, lake, river, or stream. This process is important because many chemicals will dissolve in this water as it flows. Thus, much water pollution comes from runoff originating in agricultural or industrial areas (see Chapter 15).

6. Infiltration and percolation

   i.   *Infiltration* is the process of water being absorbed into the ground. This process is essential for plant roots to uptake water.

   ii. *Percolation* is the downward movement of infiltrated water through the layers of soil, subsoil, and bedrock to the aquifer.

7. Developing land for human uses (urbanization) affects the water cycle:

   i. Infiltration and percolation are reduced because buildings and pavement cover land, so water runs off into local waterways. As a result, flooding after heavy rains is often a problem.

   ii. Runoff from agriculture, roadways, neighborhoods, and so on, may contaminate infiltrated water with pollutants like nitrates, phosphates, and toxic chemicals.

   iii. Removal of plant life leads to reduced transpiration, which contributes to a drier, often warmer microclimate in urban areas.

E. Phosphorus Cycle. The phosphorus cycle shows how atoms and molecules containing phosphorus move through various sources and sinks.

1. Rocks and sediments are the major reservoirs for phosphorus.

2. Because there is no gaseous component to the phosphorus cycle, it is often a limited factor in aquatic and terrestrial ecosystems.

3. Within a human scale of time, most phosphorus is being washed from rocks and stream beds into marine sediments (Earth's major reservoir of phosphorus). We mine sediments and land-based phosphorus to make fertilizers that increase plant growth. Runoff of these fertilizers into water may severely disrupt aquatic ecosystems because this excess nutrient often leads to algal blooms. This is one type of cultural eutrophication.

Test Tip

Questions on the AP® exam require an ability to show how the processes described in the biogeochemical cycles highlighted here relate to sustaining life on Earth.

REFERENCES

Soil Triangle: *https://study.com/academy/lesson/soil-texture-triangle-definition-use.html*

The Living World: Ecosystems

Chapter

7

I. Ecosystems Come from Biotic/Abiotic Interactions

A. Examples of Interactions

1. Because matter is infinitely recycled, there is a constant interplay as chemicals move from living things (biota) to the abiotic environment, and back into living things. Here are some examples:

    i. The availability of abiotic soil nutrients such as nitrogen, phosphorus, and potassium often limit the growth of plants within an ecosystem.

    ii. Adding nutrients to the soil or water may greatly increase plant growth. Although this may be good for growing crops, it can be devastating for aquatic ecosystems. The result can be cultural eutrophication and lead to algae blooms, lowered dissolved oxygen, and fish kills.

B. Biomes

1. A biome contains characteristic communities of plants and animals that result from, and are adapted to, its climate.

2. Earth's terrestrial biomes are determined by temperature and precipitation. This data can be represented by a climograph or climatogram.

    i. As you learned in Chapter 5, the movement of air in the Earth's convection cells combines with proximity to warm or cool ocean currents to determine an area's temperature and amount of annual precipitation.

    ii. Areas with high precipitation tend to have forests.

    iii. Lower precipitation inhibits the growth of trees. This leads to the development of grasslands, tundras, or deserts.

3. Types of biomes.

   i. Forests. Forests are areas where woody trees dominate and eventually form a closed canopy and shaded understory.

   ➤ Tropical Rainforests. Rainforests are found mostly within 25° north or south of the equator. Here are some of their defining characteristics:

   — Very high rainfall, with two seasons, wet and dry, due to yearlong warm weather.

   — Plants tend to be broad-leaved and evergreen; although rainforests contain Earth's highest biodiversity, their soils are nutrient-poor and unsuitable for farmland if the forest is cleared.

   ➤ Temperate Deciduous Forests. Temperate deciduous forests are found within the middle latitudes, between the poles and the tropics. Here are some of their defining characteristics:

   — Thirty to sixty inches of rainfall per year, distributed fairly evenly throughout the year.

   — Four distinct seasons that are roughly equal in length and primarily determined by changes in temperature: spring, summer, autumn, and winter. They are prone to frost and snow in winter.

   — Broad-leaf hardwood trees like oak, maple, hickory; shrubs like rhododendron and azalea; an understory of shade-tolerant herbaceous plants; ground mosses; and lichens.

   — Because of autumn leaf fall and decomposing leaf litter, soils tend to be well developed and rich in nutrients.

   — Locations include the eastern U.S. and central Europe, as well as eastern Asia, the southern tip of South America, New Zealand, and southeastern Australia.

   ➤ Boreal Forests (Taiga). The boreal forest is found in a broad belt across the northern latitudes from 50° to 60°. Here are some of their defining characteristics:

   — Twelve to thirty-five inches of precipitation per year, mostly in the form of snow and sleet.

- Long, dry winters and short, moderately warm, wet summers with a growing season of about 130 days.

- Because of the short growing season, trees are evergreen, bearing resin-filled, needle-like leaves.

- Tree species include spruce, pine, and fir with very little understory vegetation.

- Soils tend to be thin, acidic, and nutrient-poor.

- Locations include a broad band across the northern latitudes of North America (especially Canada and Alaska), northern Russia (Siberia), and Scandinavia.

ii. Grasslands. These areas, found in the middle latitudes, are characterized by a lack of any type of closed canopy from woody trees.

➤ Most of the vegetation is herbaceous grasses and flowers, with some shrubs and few trees. Vegetation is limited to fast-growing herbaceous plants due to the lack of rainfall (only 10 to 35 inches annually), a short growing season (100 to 175 days), and naturally occurring fires.

➤ Summers are warm and often humid; winters are cold and dry; soils tend to be thick and extremely rich in nutrients. When cleared, grasslands make excellent farmland.

➤ Notable plant and animal species include grasses, sunflowers, goldenrod, coyotes, buffalo, prairie chicken, dung beetles, and crickets.

iii. Wetlands are areas that are at least partially flooded during part of their annual cycle.

➤ Many remain flooded year-round, while others are dry for a portion of the year.

➤ Plant life in wetlands must be adapted to moist conditions and include species such as lilies, cat tails, iris, cypress, and gum trees.

➤ Wetlands have a very high biodiversity of animals. Wetlands provide valuable ecosystem services such as filtering water and minimizing flooding.

➤ Wetlands provide habitat to a wide variety of wildlife, including many endangered species.

iv. Chaparrals are arid shrub lands that are most notable for their tendency to be shaped by fire. Chaparral landscape is often used as the setting for movie westerns and is characterized by flat or rocky plains and mountain slopes. Here are some of their defining characteristics:

➤ Very low rainfall. Chaparrals are often found in the zones between deserts and grasslands. The climate is often-called *Mediterranean* because it is hot and dry for most of the year.

➤ In years of increased rainfall, fires ironically are more common due to the increased plant growth, which dries and acts as kindling for fires once the rainfall stops. There has been much controversy about the effect of fire suppression on chaparrals, with most people believing that it just increases the possibility of more severe fires later as biomass builds up.

➤ Plants and animals are adapted for conserving moisture. Reptiles have thick-scaled skin; mammals tend to be nocturnal; plants like cactus, creosote bush, manzanita, and yucca tend to have hard, thick leaves.

➤ Locations include California, the Mediterranean coast of Europe, the western cape of South Africa, and southwestern Australia.

v. Tundras are the coldest biome. They may be either at far north latitudes (Arctic tundra) or high elevations (alpine tundra). Here are some of its defining characteristics:

➤ Year-round cold and dry conditions, low growing vegetation, a short growing season, and low biodiversity.

➤ Because the growing season is short (50 to 60 days) and there is low precipitation (less than 10 inches per year), plants are very slow growing, extremely fragile, and slow to recover from damage.

➤ Animal species in the Arctic tundra include the arctic fox, ground squirrel, caribou, and ravens; in alpine tundra, mountain goats are prevalent.

vi. Deserts are characterized by extremely hot and dry conditions; deserts are found mostly near the Tropics of Cancer and Capricorn (latitude 30° north and 30° south). Here are some of their defining characteristics:

➤ Rainfall in deserts is less than 10 inches per year but may be as low as 1 inch in the driest deserts.

➤ Temperatures tend to be hot for most of the year, getting as high as 190°F in some areas.

➤ Soils tend to be sandy and fast draining, with no leaf litter or topsoil.

➤ Species include plants that are adapted to the extreme heat and drought like cactus, yucca, and agave; animals with scaled skin such as snakes, lizards, and tortoises; and burrowing animals, such as kangaroo rats and tarantulas.

vii. Aquatic ecosystems are found in the water and can be either freshwater or saltwater. (The following list includes only those aquatic systems you are likely to see on the exam.)

➤ Ponds and lakes are *inland freshwater ecosystems.*

➤ Estuary. Often found at the mouth of a river where it drains into a sea, an estuary is a partially enclosed body of water containing a mix of salt- and freshwater.

Test Tip

Coral Reefs are an underwater ecosystem composed of colonies of coral polyps held together and their calcium carbonate shells. These ecosystems are high in productivity and biodiversity. Familiarize yourself with a map of the location of different biomes and how to interpret a climate graph. Although there is a lot of information about biomes presented, this is not a topic that is worth many points on the test.

II. Resources Availability Impacts Species Interactions

A. Ecosystems can be studied from large scale to small: biosphere, ecosystem, community, population, species.

B. Symbiosis

1. Symbiosis is a close and long-term interaction between two species in an ecosystem. Types of symbiosis include mutualism, commensalism, and parasitism.

    i. *Commensalism* is a relationship between species in which one benefits and the other is unaffected. Some common types of commensalisms are housing, feeding, and transportation relationships.

> ➤ *Housing.* Many "air plants" like bromeliads, lichens, and orchids grow on the trunks and branches of trees but do not harm or benefit the tree.

    ii. *Mutualism* is a relationship between two or more species in which all members benefit directly from the interaction. Examples include the following:

> ➤ Many species of corals found on tropical reefs have a relationship with a photosynthetic algae living within their bodies. The algae perform photosynthesis and share some of the resulting glucose with the coral in exchange for a safe place to live near the water's surface.

> ➤ *Pollinators and seed dispersers.* Many flowering plants provide nectar or even nutrient-rich pollen and fruits to insects and small mammal species in exchange for the service of dispersing their seeds or pollen some distance away from the parent plant.

    iii. *Parasitism* is a relationship in which one species benefits at the expense of its host.

> ➤ Many parasites actually live on or in the bodies of larger organisms feeding on their blood or body fluids (e.g., plasmodium, tapeworms).

> ➤ Cowbirds are a type of nest parasite. They sneak in and lay their eggs in the nest of another bird, leaving the eggs to hatch and be raised by an unsuspecting host parent. Cowbird chicks tend to be larger and more aggressive than the host's own chicks and often outcompete them for food and parental attention.

C. Competition

    i. In many cases, a common resource is essential to the life of organisms of several different species. Whether it be fresh water, a food source, or space for burrows or nests, there is often a struggle for access to these resources.

> ➤ *Competitive exclusion* is the principle that two species cannot coexist if they are in direct competition for the same resource. So, no two species can exist in the same

niche at the same time. One of the species will always be slightly better adapted to exploiting the resource and will "competitively exclude the other," ultimately driving the other species either to extinction or to shift to some other resource or habitat to survive. This concept is known as Gause's law of competitive exclusion.

➤ *Exotic species* are those not originally from the place in which they are now found.

— Exotics may be introduced by humans who intentionally transport them from place to place (such as rabbits into Australia or starlings into the U.S.) or who intentionally release them into the environment (such as pythons into the Florida Everglades).

— Exotics can also be transported unintentionally as hitchhikers (such as zebra mussels in the ballast water of transoceanic ships).

— In some cases, self-directed migrations of organisms may lead them to become exotic species. Armadillos migrated from South America via Mexico into the southern U.S., and coyotes are considered invasive species in the southeastern U.S., where they have become established due to natural range expansion and human imports.

D. Different Species/Different Roles in Ecosystems

1. *Keystone species* are one or more species that are particularly important to ensuring the stability of an ecosystem. Examples include:

i. *Predators of herbivores.* Sea otters, for example, keep populations of sea urchins in check. Without this population control, sea urchins would decimate kelp plants, which are the primary producers in the kelp forest aquatic ecosystem off the Pacific coast of North America. Wolves feeding on caribou in the North American grasslands can also be considered a keystone species. This is an example of a trophic cascade, powerful indirect interactions that can control entire ecosystems.

ii. *Mutualists.* Many species have relationships in which each provides some necessary resource or service to the other. Removal of one of these species may lead to extinction of its partner species, ultimately affecting the entire ecosystem.

> ➤ *Pollinators.* A plant provides nectar as food for the polli-
> nator; the pollinator moves from plant to plant spread-
> ing pollen.

> ➤ *Seed dispersers.* Squirrels eat the acorns of oak trees and
> bury many of them to eat later. Some of the acorns are
> never recovered by the squirrel and then sprout and
> grow.

iii. *Environmental engineers.* Some species physically manipulate
the environment in such a way that the changes they intro-
duce make the ecosystem possible.

> ➤ Beavers are the ultimate environmental engineers. Their
> dam-building activities alter the flow of river water,
> creating ponds that provide habitats for a wide array of
> other species.

> ➤ Prairie dogs dig extensive networks of underground
> tunnels where they live in large colonies. Their tunnels
> channel rainwater underground, which reduces runoff
> and soil erosion.

2. *Indicator species.* Some species, by their presence or absence in an
ecosystem, can give information about the physical or chemical
characteristics of that environment.

i. *General environmental characteristics.* By looking at what species are present in an area, one may learn about the characteristics of that area.

> ➤ The calamine violet will grow only in soils that have
> high concentrations of zinc.

> ➤ Macroinvertebrates like mayflies, caddisflies, stoneflies
> are useful indicators of water quality in aquatic eco-
> systems. The larval stages of these insects are able to
> survive only in water with high dissolved oxygen, which
> usually correlates with high water quality.

ii. *Pollution.* Some species are very sensitive to the presence of
pollutants.

> ➤ Many species of frogs have recently had a dramatic
> increase in physical defects, such as extra or missing
> legs. These defects have been linked, in some cases, to
> pollutants in the water the frogs inhabit. Amphibians in
> general tend to be sensitive to changes in water quality.
> Their thin permeable skin allows many pollutant chemi-
> cals from the water to enter their bodies.

➤ Lichens are an example of air quality indicator species. There are three classes of lichen: crustose, foliose, and fruticose (named by their appearance: crusty, leafy, or bearing fruits). Crustose are the most resistant to air pollutants, foliose are more sensitive to air pollutants, and fruticose are the most sensitive over all. By looking at the distributions of these three types of lichen, one can determine some basic information about how clean or polluted the air is in a location.

iii. *Environmental change.* Some species are more sensitive to changes in the environment. For example, most coral species that populate tropical coral reefs can tolerate water within a very narrow temperature range. Many scientists are closely monitoring coral reef health as a possible indicator of global warming.

3. *Invasive species.* Some species can move into a new territory and beat out the native species in the competition for resources. Invasive species can also be referred to as exotic or non-native.

i. When organisms are moved from their endemic or native habitat to a new one, they may establish a population if conditions are within their range of tolerance. In the new habitat, these newcomers are called an *exotic species*.

ii. In some cases, there will be no natural predators, diseases, or competitors to keep the exotic population in check. Exotic species that adapt very well to their new environment are referred to as *invasive* if they alter their newly adopted ecosystem.

iii. Invasive species are usually able to reproduce quickly and thus leave large numbers of offspring, are very tolerant of environmental changes, and tend to be generalist species able to survive off a wide variety of resources.

iv. Examples include plants like Kudzu and Brazilian pepper and animals like the German cockroach, European starling, cane toad, and even feral cats.

E. Energy Transfer in Ecosystems

1. Photosynthesis

i. *Photosynthesis* is the process by which green plants use energy from the sun to make glucose (food). In the process, plants take in carbon dioxide (CO_2) and release oxygen (O_2) as a waste product.

ii. The chemical formula for the photosynthesis reaction is:

$$6CO_2 + 6H_2O \rightarrow C_6H_{12}O_6 + 6O_2$$

iii. Photosynthesis by plants is significant because it fixes carbon into a usable form and produces the oxygen that makes life on Earth possible.

iv. Because it uses carbon dioxide as a reactant, photosynthesis is also responsible for helping to maintain the carbon dioxide balance in the atmosphere (carbon dioxide is an important greenhouse gas).

Test Tip

Recall that photosynthesis and respiration are processes in the carbon cycle that move carbon from one reservoir to another quickly. This is an example of how content is spiraled from one unit to another. You should develop a deeper understanding of the carbon cycle as you use that knowledge to explain energy transfer in ecosystems.

2. Cellular Respiration

   i. This process is used by aerobic organisms to break down sugar molecules and convert them into energy. All cells (even those of the plants that made the sugar) undergo cellular respiration to unlock the energy contained in their food.

3. Anaerobic Respiration-Fermentation

   i. Some prokaryotes and eukaryotes use anaerobic respiration in which they can create energy in the absence of oxygen.

F. Productivity in Ecosystems

1. Primary Productivity

   i. *Primary productivity* is the rate at which solar energy is converted into organic compounds (sugars) through photosynthesis.

   ➤ Gross primary productivity (GPP) is the total amount of energy from the sun that plants convert to organic matter.

   ➤ Respiration (RESP) is how much of this energy plants use for their own growth and metabolism (through the process of cellular respiration).

> Net primary productivity (NPP) is how much of this energy is then available to the consumers who get their energy by eating plants.

> Calculating primary productivity. Productivity is measured in units of energy per unit area per unit time (e.g., $kcal/m^2/yr$).

Gross primary productivity = net primary productivity
+ cellular respiration

> Photosynthetic rates vary by depth because most red light is absorbed in the top one meter of water and blue light can penetrate more deeply.

ii. Lab experiment. A common lab experiment for environmental science classes is to calculate the GPP in an ecosystem by measuring oxygen production by plants undergoing photosynthesis and cellular respiration (NPP), then adding back the amount of oxygen consumed by respiration by measuring oxygen loss in plants kept in the dark (RESP).

Math Practice

Dr. Lawrence has set up an experiment in her lab to determine the gross primary productivity of her fish-tank ecosystem. She first measures the net primary productivity to be 5,000 kcal/m^2 per year. She then measures the respiration by aquatic producers to be 7,000 kcal/m^2 per year. What would her calculated gross primary productivity be?

ANSWER:

To calculate the gross primary productivity of an ecosystem, use this equation:

GPP = NPP + RESP

NPP = 5,000 kcal/m^2 per year

RESP = 7,000 kcal/m^2 per year

GPP = 5,000 kcal/m^2 per year + 7,000 kcal/m^2 per year
= 12,000 kcal/m^2 per year

G. Trophic Level

1. Recall how *matter* cycles through biogeochemical cycles. Ecosystems also depend on a continuous flow of high-quality *energy* to function.

2. Food Chain

   i.   A food chain is a simplified visual representation of the organisms that feed on one another in an ecosystem. For example:

      Grass → Cricket → Shrew → Snake → Hawk

3. Food Webs

   i.   Because of the complexity of feeding behaviors in a typical ecosystem, a food web is used to represent the many food chain possibilities.

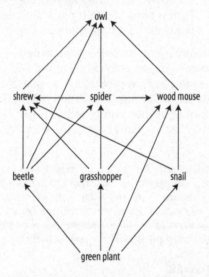

You may be asked to draw a food web from a description of an ecosystem. Remember that, when interpreting or drawing a food web, the arrow is always drawn from food organism to eater, in the direction that energy is moving (from prey to predator). Also remember that you need to draw an arrow from each organism to every other organism that might consume it. Several arrows may be needed from each organism.

4. Feeding Levels

    i.   *Trophic levels* are feeding levels determined by an organism's position in a food chain. The typical trophic levels in an ecosystem are:

      ➤ Producers. Known as autotrophs or self feeders, these are mostly green plants that produce their own food (sugars) through photosynthesis.

      ➤ Consumers. Also known as heterotrophs, these are organisms that must eat other organisms to survive. They can be primary consumers that eat plants or secondary consumers that eat an animal that ate a plant. This relationship is represented in food webs and food chains.

      ➤ Decomposers. This class includes all organisms that feed on dead plant or animal materials. Some classes of decomposers include detrivores (eating dead plant materials) and scavengers (eating dead animal carcasses).

5. Efficiency of Energy Conversions

    i.   As energy moves through a food chain, from one trophic level to the next, much of the energy is lost in the process. This conversion is often represented as an energy pyramid (E.U. stands for "energy units"):

| Tertiary consumers (carnivores) 10 E.U. |
| Secondary consumers (omnivores) 100 E.U. |
| Primary consumers (herbivores) 1,000 E.U. |
| Producers (autotrophs like plants and photosynthetic algae) 10,000 E.U. |

The sun
(the ultimate source of energy in most ecosystems)
1,000,000 E.U.

   ii.   Notice that the efficiency of the conversion of energy from sunlight to producers is approximately 1 percent.

   iii. The approximate efficiency of the conversion of energy at each step after this (producer to primary consumer and up) is 10 percent.

   iv. What happens to all the lost energy?

      ➤ Some is used by the organism itself for growth and other activities of life.

      ➤ Some is lost to the environment in the form of heat.

   v.   This explains the following:

      ➤ Rarely does an ecosystem have more than four or five trophic levels represented. Beyond that point, there is very little energy left to support life.

      ➤ Many animals eat a variety of food types. For example, when you eat a salad, you are acting as a primary consumer at a lower trophic level than when you eat a steak and thus act as a secondary consumer.

      ➤ It costs more to produce and buy a pound of meat than a pound of corn.

      ➤ Many more people can be supported as vegans than as omnivores using a given area of land.

Math Practice

As part of his graduate research, Mark Waters has measured the amount of energy at the producer level of a grasslands ecosystem to be 720,000 kcal/m² per year. Now he needs to calculate the approximate amount of solar energy coming into this ecosystem. He would also like to calculate the amount of energy available at each trophic level in this ecosystem. He has asked you to show him how to do these calculations. Can you help?

(continued)

ANSWER:

To calculate the solar energy, remember that the energy at the producer level is approximately 1 percent of the incoming solar energy, so do the following:

1. Convert 1 percent to its decimal form = 0.01

2. Use the equation:

    Solar Energy(0.01) = Producer Energy

3. Manipulate the equation to:

$$\text{Solar Energy} = \frac{\text{Producer Energy}}{0.01}$$

4. Insert your numbers into the equation:

$$\text{Solar Energy} = \frac{720,000 \text{ kcal/m}^2 \text{ per year}}{0.01}$$
$$= 72,000,000 \text{ kcal/m}^2 \text{ per year}$$

To calculate the energy available at each of the trophic levels above producer, multiply the value of any trophic level by 10 percent (or 0.1) to determine the approximate energy available at the next level up:

Producers × 0.1 = Primary Consumers

(720,000 kcal/m^2 per year × 0.1 = 72,000 kcal/m^2 per year)

Primary Consumers × 0.1 = Secondary Consumers

(72,000 kcal/m^2 per year × 0.1 = 7,200 kcal/m^2 per year)

Secondary Consumers × 0.1 = Tertiary Consumers

(7,200 kcal/m^2 per year × 0.1 = 720 kcal/m^2 per year)

REFERENCES

Food web diagram from Field Studies Council—Urban Ecosystems: *https://www.invertebrate-challenge.org.uk/urbaneco/urbaneco/ introduction/feeding.htm*

The Living World: Biodiversity

I. Introduction to Biodiversity

A. Diversity

1. The diversity of an ecosystem can include species, genetic, and habitat diversity.

B. Ecosystems with High Biodiversity

1. Species richness is a count of the number of different species. For example, if you survey a grassy area and discover a species of mouse, two cricket species, four species of flowers, and three different grass species, species richness would equal 10 species. Ecosystems with a high species richness are more stable and resilient.

   i. Species evenness describes the relative abundance of each species represented. For example, a system with 90 of each cricket species, 110 of each species of flowers, 100 of each grass species, and so on, would have much higher species evenness than one in which there were 40 of each cricket, 100 of each flower, and 400 of each grass.

   ii. Genetic variation within a species. A population bottleneck can lead to a reduction in genetic diversity; a genetically diverse population can better respond to environmental change.

      ➤ Bottleneck. When a species is reduced to a small number of individuals, the genetic diversity of that species is naturally reduced. If later events allow the species to rebound, there will still be a reduced genetic diversity because all genetic stock can be traced back to the small number of individuals. This leaves the species more vulnerable to extinction due to environmental changes, disease, and so on.

C. Biodiversity by Biomes/Map Regions

1. In general, areas of highest biodiversity are found near the equator; tropical rainforests, estuaries, and coral reefs all tend to have high diversity. This is likely due to the stable, warm temperatures and relatively abundant supply of water.

2. Conversely, areas with great fluctuations in temperature, or harsh climates that are very cold and/or dry, with low nutrient levels, tend to have much lower diversity. This includes the polar ice caps, the Arctic tundra, and open ocean zones.

Test Tip

The AP® exam commonly has a set of questions in which you will have to identify regions on a map. As you study topics like biomes and climate, human population growth, and land and water use, be sure to pull out a world map to see where events are happening. Make sure that you recognize all of the continents and major regions and countries (the U.S., China, India, the Middle East, etc.). This knowledge will help you interpret the map questions.

D. Island Biogeography

1. Island biogeography is the scientific theory developed to explain the differences in biodiversity among islands. It involves an interaction between the immigration of new species to the island and the extinction of species already there. In general, the theory states the following:

    i. Highest biodiversity corresponds with:

       ➤ large islands due to their tendency to have a wider variety of habitats and resources available to support immigrant species.

       ➤ islands closer to the mainland because it is easier to migrate across a small distance and survive.

    ii. Conversely, low biodiversity is expected on smaller islands that are farther away from the mainland.

    iii. In addition to explaining actual islands surrounded by water, island biogeography is also useful in understanding habitat islands.

       ➤ Habitat islands are isolated ecosystems—for example, a patch of protected forest surrounded by human development.

➤ Island biogeography theory has led to better management of habitat islands like national parks and state forests, including the use of corridors to encourage movement of species from one "island" to another. Because of limited resources, island species tend to be specialists and thus susceptible to invasive species, which tend to be generalists.

Test Tip

Review the task verbs used on the FRQs in Chapter 23. A typical question might ask you to identify why islands are prone to invasive species, explain characteristics of successful invaders, and make a claim about the relationship between island size and biodiversity.

E. Biodiversity as an Indicator of Disturbance

   1. To achieve the highest biodiversity, just the right amount of disturbance is needed.

➤ As we measure biodiversity of a system over a period, a decrease is often an indicator of some type of negative disturbance. Pollution, habitat loss, and the introduction of exotic species are all examples of disturbances that may result in lowering the biodiversity of an ecosystem.

➤ Ecosystem disturbance is not always bad for biodiversity. The highest levels of biodiversity are often found in ecosystems where disturbances like wildfires or tropical storms occur. This stimulates competition and allows both r-selected and K-selected species to thrive.

F. Natural Selection and Survival

   1. Natural Selection

     i. Organisms that are best adapted to survive and reproduce in their given environment tend to thrive and produce more offspring. As a result, the *alleles* (genetic information) possessed by these successful organisms become more common in the population over time.

      ii.   Natural selection is often stated as survival of the fittest. This works as long as you know that "reproductive fitness" is the intent, meaning that an individual leaves many offspring. Fitness does *not* have to mean strength, intelligence, and so on. It could be as simple as being the perfect shade of green to blend in with the surrounding environment.

  2. Adaptation

      i.   An *adaptation* is any trait giving an organism a survival or reproductive advantage.

  3. Genetic Resistance

      i.   *Genetic resistance* is an adaptation in which members of a species become able to tolerate some usually fatal condition and pass that resistance on to their offspring. Examples include:

> ➤ Antibiotic resistance. Staphylococcus bacteria has become increasingly resistant to the antibiotics used to treat staph infections.

> ➤ Pesticide resistance. Pest species have developed a genetic resistance to the pesticides used to control them.

> ➤ Resistance to disease. For example, people who are heterozygous for the sickle cell anemia trait have a genetic resistance to malaria due to having a proportion of misshapen red blood cells. This makes their bodies inhospitable to the reproduction of the plasmodium parasite that causes malaria.

  G. Ecosystem Services

  1. Ecosystem services include all of the natural processes that contribute to one's ability to live on Earth. Some examples include:

      i.   provision of animals and plants for food, clean water, fertile soil, and trees for shelter.

      ii.   regulation of Earth's climate, flood control, filtration, and purification of waters.

      iii.   supporting human activities through the cycling of matter, decomposition of wastes, and pollination of plants.

      iv.   cultural services like providing spiritual and aesthetic support and recreation opportunities.

2. Some human activities that compromise nature's ecosystem services include the introduction of non-native species, pollution, deforestation, and development of land to accommodate a growing population.

II. How Ecosystems Change

A. Impact of Natural Disruptions

1. Earth's climate has changed over geologic time.

2. Over geologic time, there have been varying amounts of glacial ice on Earth, which in turn has altered sea levels.

3. Animals migrate for various reasons including natural disruptions.

4. The range of abiotic conditions such as temperature and pH over which an organism can thrive is known as *ecological tolerance*.

B. Ecological Succession

1. An example of the changes that take place in ecosystems is *ecological succession*. This is the process in which a new or disturbed system moves from a state of very little life to a fully developed ecosystem.

   i. Primary succession. Starting from bare rock, slowly over time, soil is built up and then plants and animals colonize. Drastic disturbances like volcanic eruptions or major mudslides, as well as new island formations, can result in primary succession.

   ii. Secondary succession. Disturbances like forest fires, deforestation, and hurricanes may remove most of the organisms in a system but leave behind soil and seeds. The process of nature reclaiming an ecosystem that used to exist but was disturbed is secondary succession.

   iii. Pioneer species are the first species to colonize an area during ecological succession. During primary succession, the most common pioneers are lichens and small ferns and mosses that grow directly on rock. During secondary succession, grasses and other fast-growing herbaceous species are often the first pioneers.

   iv. Disturbing an ecosystem will affect the total biomass and species diversity.

III. Population Change Over Time

A. Population Growth and Resource Availability

1. Resource availability limits population growth. When resources are plentiful, populations will grow. When resources become limited, populations may experience high death rates and lower survivorship.

   i. Linear growth is a state where a population grows by a *fixed amount* over time. On a graph, it is a straight, diagonal line.

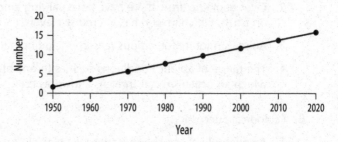

Linear Population Growth By Decade

   ii. Exponential growth is a state where a population grows by a *fixed percentage* over time. Exponential curves are shaped like the letter J and are characterized by slow growth at first, followed by an explosion of very rapid growth. This type of growth is characteristic of r-selected species.

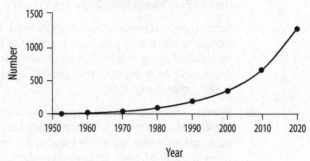

Exponential Growth By Decade

iii. Logistic growth. Most populations show this type of growth. On a graph, this curve is S-shaped. Logistic growth is characterized by initial exponential growth followed by a leveling off of the curve over time. This type of growth is characteristic of K-selected species.

Logistic Growth By Decade

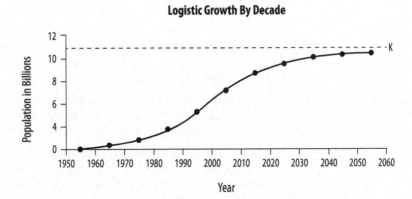

iv. Carrying capacity is defined as the maximum number of individuals that can be supported indefinitely in a population. It is determined by the abundance of resources in a given environment.

➤ As a population continues to grow, eventually the rate of growth will be kept in check by environmental resistance. For example, food shortages lead to competition and starvation, space becomes limited, and so on.

➤ If a population grows far beyond its carrying capacity (overshoot), the result may be so damaging to the environment that the overall carrying capacity is lowered for future generations.

➤ On a graph, carrying capacity is often shown as a line marked with the letter K and is usually the point in a logistic growth curve where the population growth slows or stops. On the graph above, the line representing K is drawn just above the 10 billion mark.

2. r- and K-selected species

i. A species may be classified by what type of population growth it usually experiences based on the life history of the species.

➤ r-selected species tend to be species with high biotic potential and rapid population growth rates. These species often show the following traits:

— Small body size, short life span, short time to reach sexual maturity, and short gestation period (time from conception to birth).

— Little parental investment in each offspring, many offspring born per litter or nest, and frequent reproductive events.

— Examples of r-selected species include small rodents such as mice and rats, insects such as aphids, and plant species such as dandelions.

➤ K-selected species tend to have a much lower biotic potential with more stable population sizes that hover closely around the carrying capacity (K). Common traits include the following:

— Larger body size, longer life span, usually a year or more to reach sexual maturity, and a longer gestation period (months rather than days or weeks).

— Greater parental investment, with parents often spending a great deal of time and energy protecting and teaching their offspring for a year or more; fewer offspring per reproductive event (often only one or two); breeding less frequently (often on an annual cycle).

— Examples of K-selected species include elephants, humpback whales, and humans.

➤ Not all species fit neatly into one category. For example, green sea turtles lay several hundred eggs in a nesting season, show no parental care to offspring, and have very high hatchling mortality rates like an r-selected species. But they also have a life span of many decades, take about twenty years to reach sexual maturity, usually breed only every second year, and can weigh hundreds of pounds. These are all characteristics common to K-selected species.

3. Generalists and Specialists. Generalists take advantage of rapidly changing habitats. Specialists compete well when conditions are fairly stable.

B. Survivorship Curves

1. Another way to classify populations is to look at how many individuals survive at each phase of life. There are three types of survivorship curves.

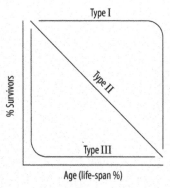

i. Type I. These species have a high survival rate throughout most of their life span.

➤ Most die at an old age.

➤ Because there is usually much parental investment in protecting and caring for their young, each offspring has a better chance to survive into adulthood. This is typical of K-selected species.

➤ Humans living in more developed countries, zoo animals, and pets also fall into this category.

ii. Type II. These species have a relatively constant death rate throughout their life span.

➤ They are equally likely to die at any time in their life.

➤ Prey species such as squirrels and small rodents, songbirds, and many reptiles show this type of survivorship.

iii. Type III. These species show a high death rate at the beginning of their life span, which usually levels off once adulthood is reached.

➤ Many r-selected species show this pattern of having many offspring with the expectation that very few will survive.

➤ Examples include plant species that produce many seeds, oysters, and sea turtles.

PART III

HOW HUMANS USE AND CHANGE EARTH

Populations: Human Population Dynamics

Chapter

9

I. Introduction to Population Demographics

A. World Population

1. Human population growth is impacted by cultural and social factors.

2. The world's population is approximately 7.8 billion (2020).

3. The global population growth rate is 1.05 percent (2020).

B. Most Populous Countries (2020)

1. China

2. India

3. United States

C. Total Fertility Rate

1. Total fertility rate (TFR) is the number of live births, on average, that each woman of a nation will have during her life. It is influenced by the age at which females have their first child, educational and employment opportunities for females, and family planning access and policies.

2. The countries with the lowest TFR also have the lowest population growth.

3. Replacement level fertility equals the TFR necessary to keep population numbers stable. In more developed countries, the replacement level fertility is about 2.1 (taking into account low levels of infant mortality). In countries with higher mortality rates, replacement-level fertility is higher.

4. In general, TFR tends to be lowest in the countries with low infant mortality rates and the highest standards of living.

Test Tip *Sharpen your quantitative skills in AP® Environmental Science by solving sample AP® problems.*

D. Doubling Time

1. When we study population growth, a common statistic given is *doubling time*. This is the number of years it will take for the population to grow to twice its current size.

2. The rule of 70 is an equation that calculates doubling time by dividing 70 by the percentage population growth.

$$\text{Doubling time} = \frac{70}{\text{percent population growth}}$$

3. From the same equation, the percentage population growth can be calculated by dividing 70 by the doubling time.

$$\text{Percentage population growth} = \frac{70}{\text{doubling time}}$$

E. Natural Annual Rate of Increase

1. Natural annual rate of increase is calculated using the difference between the crude birth rate and the crude death rate for a population.

2. Crude birth rate is the number of childbirths per 1,000 people in any given year.

3. Crude death rate is the number of deaths per 1,000 people in any given year.

F. Growth Rate and Technological Advances

1. Growth rate in the last 100 years has been greatly affected by technological advances.

2. The Industrial Revolution and the subsequent agricultural and medical advances first led to a population boom as life spans increased and food supply became more stable.

3. In the twenty-first century, people in nations with the greatest access to technology are choosing to have fewer children, resulting in decreasing rates of population growth.

II. Social and Environmental Problems Related to Population Growth and Distribution

A. World Hunger

1. According to most estimates, there is enough food produced on a global scale to feed all of Earth's population. However, the food supply is not distributed equally among all people.

2. The Earth's carrying capacity limits human population growth based on the Malthusian theory. Populations grow geometrically, while food supplies grow arithmetically; therefore population will exceed food supply.

3. Density independent factors (fire, drought, severe storms) as well as density dependent factors (disease, clean water, food) affect population growth.

B. Pathogens and Infectious Disease

1. As population growth leads to more people living closer together (increased density), the opportunities for outbreaks of disease dramatically increase. Pathogens take advantage of environmental change to infect human populations.

   i. A changing climate allows pathogens and vectors to spread north and south from equatorial regions, infecting more people.

   ii. Deforestation is bringing humans into contact with pathogens and vectors they have no resistance to.

2. Some diseases require a vector, a "middleman," that carries the infection from one organism to another. A vector is an organism,

usually a biting insect, that transmits a disease from one animal to another. Vector-borne diseases cannot be passed directly from one person to another. Examples include the following:

    i.   Malaria is caused by a blood-borne parasite called *Plasmodium*. It is transmitted by the female *Anopheles* mosquito feeding on the blood of an infected person, then passing on the parasite to the next person from whom they feed.

    ii. Bubonic plague is transmitted from one person to another via fleas.

    iii. Ticks are common vectors for various diseases, including bacterial infections, chemical toxins, parasites, and viruses. Lyme disease is one notable bacterial disease transmitted by the bite of an infected tick.

3. Some older diseases, thought to be totally eradicated are returning due to population growth, complacency about vaccination, and antibiotic resistance:

    i.   Tuberculosis is a bacterial infection of the lungs.

        ➤ TB can be spread through the air by tiny particles from the cough or sneeze of an infected person.

        ➤ Another form of tuberculosis is spread through contaminated milk. Pasteurization of milk kills the bacteria before it can cause the disease.

    ii. Malaria is caused by a vector-borne parasite; it's common in wet tropical areas near the equator.

        ➤ The best way to control malaria is by eliminating the mosquito vector.

        ➤ DDT was and still is commonly used to control mosquitoes, but has been banned in the U.S. and other countries due to its harmful ecological effects.

    iii. Cholera is a bacterial infection that commonly spreads through an unsanitary water supply.

4. Emerging pathogens include newly discovered parasites, bacteria, and virus strains that cause a host of "new" diseases including:

    i.   HIV/AIDS was discovered in the 1980s. Although not exactly new, the human immunodeficiency virus (HIV) is still a recent pathogen in evolutionary terms.

        ➤ A person infected with HIV can be symptom-free for several years before becoming ill with acquired immune deficiency syndrome (AIDS). This makes the virus difficult to eradicate because people may be unaware that they are infected and can unknowingly spread the disease through sexual contact or the sharing of blood.

ii. H1N1 virus (swine flu) was first detected in the U.S. in 2009 and quickly spread worldwide. This virus was even more deadly because of misinformation about how it was transmitted (some people erroneously believed eating pork was the cause).

iii. SARS (severe acute respiratory syndrome) was caused by a coronavirus. It spread through close personal contact, either by inhaling airborne droplets from a cough or sneeze of an infected person, or through touching a contaminated surface and introducing the virus into the body.

➤ In 2003, SARS was first documented in Asia and rapidly became a pandemic, affecting more than 8,000 people throughout North and South America, Europe, and Asia.

iv. Zika is a virus spread by mosquitoes and through sexual contact.

v. COVID-19

➤ Coronavirus disease 2019 is a respiratory illness that can spread from person to person.

➤ The virus that causes COVID-19 is a novel coronavirus that was first identified in 2019.

➤ By mid-2020 COVID-19 had spread world-wide causing over 400,000 deaths, according to data collected by Johns Hopkins University.

vi. West Nile Virus, MERS, and cholera are also emerging infectious diseases.

5. Currently, the major public health factor is that human populations are no longer geographically isolated. Because of the increase in air travel, disease outbreaks that were formerly contained to a single group of people now have a greater potential for reaching pandemic status.

III. Resource Use and Habitat Destruction

A. Two major aspects of population growth make it an important factor in all other environmental problems: the increasing world population and increasing per capita resource use.

1. As there are more people on Earth, if per capita resource use remains constant, every additional person puts additional demand on our already limited resources. This is especially true of fresh water and arable land.

2. In reality, per capita resource use is not remaining stable. As more nations continue to reach higher levels of economic development, their people's standard of living is also increasing. So not only is Earth's population growing, each individual is also consuming more and more resources with each passing generation.

B. "The Tragedy of the Commons"

1. Garrett Hardin and the Commons

   i.   In 1968, Garrett Hardin published an article in the journal *Science* called "The Tragedy of the Commons." His idea was that there is a finite maximum number of people the Earth can support and how we use Earth's resources is an important factor in determining how many people can be sustainably supported.

   ii. "The commons" Hardin refers to is any resource used by several people (or groups) but owned and controlled by no one.

   iii. The tragedy is that each person with access to the resource has no reason to limit consumption of the resource, especially if there is no strict regulation with consequences for misuse. All it takes is one person or group to take more than his or her share to ruin the sustainability of "the commons" for all.

   iv. As a result, each user is motivated to get what he or she can while it lasts before others take what's left.

2. Examples of Resources That Are "Commons"

   i.   International Tuna Fisheries

   ➤ Bluefin tuna are extremely valuable fish. Because they migrate thousands of miles and are caught in the open ocean, no single country "owns" the bluefin tuna population. If the U.S. regulates its catch of the tuna and another country chooses not to regulate its catch, and thus overharvests the tuna, the tuna will eventually become extinct. The U.S., however, will have reduced its capacity for profit, allowing countries without any regulations to profit by overfishing the tuna population.

      ii.   Overgrazing on Public Land

          ➤ The U.S. Bureau of Land Management oversees huge tracts of prairie land throughout the U.S. that is used to graze livestock. Much of this land is leased to ranchers for grazing large herds of cattle. Early in the twentieth century, when the land was originally settled, ranchers were encouraged to maximize their profits by keeping more and more cattle on their leased land. As cattle densities increased, the prairie ecosystem was damaged, and this damage led to erosion and desertification.

      iii. Water Supply

          ➤ Aquifers (groundwater), large lakes, and rivers are all sources of clean fresh water for drinking, irrigation, and industrial uses. We all depend on these resources, but they can be damaged easily or depleted by any single party who uses them unsustainably.

      iv. More Common Resources

          ➤ Other examples include atmosphere/air, national forests and national parks, estuaries and wetlands, and the continent of Antarctica (which is managed by a consortium of nations that share land rights).

Test Tip

"The Tragedy of the Commons" is tested on every exam, usually as a multiple-choice question posing a variety of situations and asking which one of the answer choices is an example of tragedy of the commons. Remember to look for the situation of a publicly owned resource, something like the open ocean or the atmosphere, that can't be owned, is used by many, and is prone to overuse or exploitation.

C. Age Structure Diagrams

    1. Also called population pyramids, age structure diagrams show the number of people at each age group in a population. They tend to be divided with males on one side and females on the other.

    2. The diagram also shows the relative percent of the population in different age groups; pre-reproductive, reproductive, and post-reproductive.

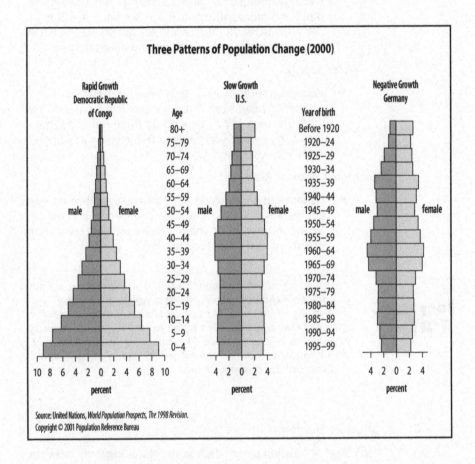

Three Patterns of Population Change (2000)

Rapid Growth
Democratic Republic
of Congo

Slow Growth
U.S.

Negative Growth
Germany

Source: United Nations, *World Population Prospects, The 1998 Revision.*
Copyright © 2001 Population Reference Bureau

D. Demographic Transition

    1. Demographic transition is the process in which the population in an LEDC (less economically developed country) starts to reflect the developmental changes taking place as that country moves toward a state of greater economic development and stability.

2. Most graphs show demographic transition occurring in four stages. Below are two ways to visualize the changes taking place: by demographic pyramids and a demographic transition birth rate/death rate graph.

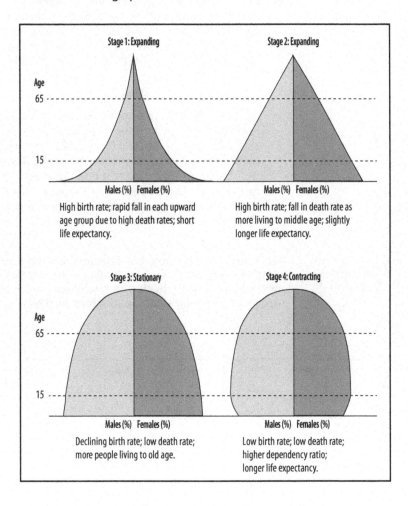

i. The diagram above shows the stages of demographic transition in age structure pyramids.

➤ Stage 1: Shows the typical pyramid for an LEDC, with a high birth rate and high death rate at all age cohorts, and short life expectancy.

➤ Stage 2: As a nation becomes more economically developed, death rates fall. Due in part to better health care, improved sanitation, and more stable diets, infant mortality decreases and life expectancy begins to increase. The birth rate is still high.

➤ Stage 3: As the nation continues to progress in its economic development, further stability, improved education, and improved medical care lead to a decrease in birth rates. At the same time, more people are living to older ages because of the reproductive time lag caused by longer human life spans; the population continues to grow even as birth rates decrease.

➤ Stage 4: The birth rate continues to decrease and then levels off with higher levels of education and health-care availability, especially for women. The population continues to have an older mean age. Eventually, this stabilizes population to a point of much slower or zero growth. Late in stage 4, after many decades of a stabilized economy and high living standards, countries begin to see declining population growth.

The following graph shows the four steps of demographic transition in terms of changing birth and death rates and their effect on population growth. This type of graph is commonly used in AP® Environmental Science exam questions.

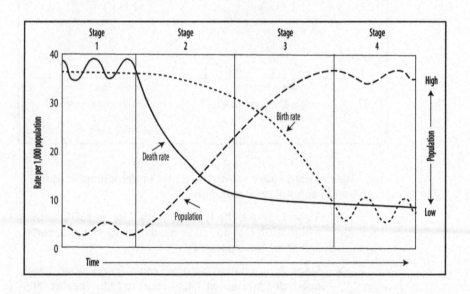

Math Practice

POPULATION CALCULATIONS

A. Rule of 70:

1. *The U.S. has a population growth rate of approximately 1 percent. In how many years will the population double if that growth rate remains constant?*

2. *Compare the doubling time of the U.S. population with that of Belize, which has a growth rate of about 2 percent.*

3. *At the current rate of population growth, Earth's population will double in about 64 years. What is the current percentage of the population growth rate?*

B. Growth rate (account for immigration and emigration)

1. *If a population has a crude birth rate of 15 per 1,000 people, and a crude death rate of 3 per 1,000 people, what is the natural annual percentage increase of its population?*

2. *Earth's population in 2011 was almost 7 billion and was growing at an annual rate of 1.1 percent. At this rate of growth, how many people would have been expected to be added in the following 12 months?*

ANSWERS:

A. Rule of 70:

1. $\dfrac{70}{1\%}$ = 70 years to double

2. $\dfrac{70}{2\%}$ = 35 years to double. The population of Belize would double twice as fast as the U.S.

3. $\dfrac{70}{64 \text{ years}}$ = 1.1% annual population growth rate

B. Growth Rate

1. Crude birth rate – crude death rate = 15 per 1,000 – 3 per 1,000 = 12 per 1,000. Adjust the decimal to reflect 1.2 per 100, or a 1.2 percent increase.

   ➤ If the crude death rate had been greater than the crude birth rate, the percentage would be a negative number (to reflect that the population would actually be getting smaller).

2. 7 billion = 7×10^9

   1.1% = 0.011

   $7 \times 10^9 \times 0.011 = 0.077 \times 10^9 = 7.7 \times 10^7$

Test Tip

It is almost guaranteed that there will be a growth-rate calculation on the exam. This is good news because they are simple as long as you are familiar with, and remember, the rule of 70!

REFERENCES

Data for population numbers, growth rates, GDP, and GNP from: U.S. Census Bureau *http://www.census.gov*

The World Factbook. CIA. *https://www.cia.gov/library/publications*

Demographic transition pyramids model from: *https://commons. wikimedia.org/wiki/File:Dtm_pyramids.png*

Demographic transition diagram from the Lewis Historical Society: *http://www.lewishistoricalsociety.com*

Earth Systems and Resources: Food Production

I. Agriculture

A. Global Food Production

1. As human population continues to grow we will need to find ways to increase food production while the Earth systems that produce food, cropland, grazing land, and ocean fisheries are being threatened.

B. Green Revolution

1. The green revolution introduced new, high-yielding varieties of wheat, corn, and rice into less developed nations like Mexico, India, China, and several others throughout Africa and Asia.

2. Some of the strategies that increased yield during the green revolution include modern equipment, genetically modified organisms (GMOs,) irrigation techniques, and the use of pesticides. All of these strategies have increased yield but may have environmental and human health impacts.

3. Genetically Modified Organisms (GMOs)

    i. **Pros:** GMO crops can often be grown in conditions that would otherwise make it impossible for the crop to grow. Varieties may be adapted to drought, salty soil, or low nutrients. GMOs may be resistant to herbicides, thus allowing for easier elimination of weedy competitors. They may also possess other desirable qualities—like golden rice genetically engineered to contain vitamin A.

    ii. **Cons:** Some scientists fear that genetically modified organisms are risky because the crop plants may accidentally be released into natural ecosystems and many of their modified traits may lead them to become aggressive exotics. There is public debate about the use of GMOs.

Free-response questions will often ask you to identify pros and cons of a practice, such as extensive irrigation from an economic, ecological, and human health perspective. Be sure you can clearly describe the impact and consequence caused by a certain practice.

C. Sustainable Food Production

1. Vegetarian Diets

   i. **Pros:** If you recall trophic pyramids and energy moving through ecosystems, you know that eating lower on the trophic pyramid is dramatically more efficient; on average, only about 10 percent of the energy from one trophic level is available to organisms on the next higher level. Therefore, a vegetarian diet or a diet low in meat is more sustainable than a meat-based diet in terms of calories that can be produced per acre.

   ii. **Cons:** The less variety in one's diet, the more prone one is to malnutrition caused by deficiencies of vitamins, minerals, and proteins. Protein deficiency diseases like *kwashiorkor* (causes puffiness and large bloated belly) and *marasmus* (severe loss of muscle tissue) are much more common in areas where diets consist mainly of a single grain staple, like corn or rice, and very little meat.

2. Eating Local

   i. **Pros:** Eating food that is grown locally is better for both the environment and the local economy. Not only is your money supporting local farmers, it is easier to obtain information about which agricultural techniques were used. Also, less fossil fuels are used to transport the food from origin to point of sale.

   ii. **Cons:** There are fewer options when one is limited by the types of foods that can be produced locally. In the northern climates during winter months, many of the fruits and vegetables that are taken for granted would not be available. In large, densely populated areas such as Chicago and New York City, it would be difficult or impossible to produce enough locally grown/raised food to supply everyone.

II. Pest Control

A. Definition

1. Pest control involves any action humans take to eliminate unwanted organisms. Although the use of pesticides has increased crop yield, it has led to human exposure to toxins and pesticide-resistant crops.

2. The pesticide treadmill is the repeated cycle of pesticide application and development of resistant populations of pests requiring more chemicals be used even as fewer pests are being killed.

3. Integrated pest management (IPM) incorporates physical, biological, and limited chemical strategies to minimize crop loss to an economically tolerable level.

B. Pros and Cons of Pest Control Methods

1. Chemical Control

   i. **Pros:** Chemicals are often highly effective in killing the target organisms. Results may be quick and dramatic and may be the only way to control a major infestation of pests.

   ii. **Cons:** Chemicals may be toxic to nontarget organisms such as children and family pets. They may also kill natural predators of the target organism. In addition, they may be harmful to the environment, especially if broadcast sprayed.

2. Biological Control

   i. Biological control is the introduction of natural enemies to deal with pest species. These enemies may be predators, parasites/parasitoids, or pathogens.

   ii. **Pros:** When it works, biological control is impressive! There is an initial expense of collecting, transporting, and releasing the enemy species. Once they are established, however, very little maintenance is required. Pests are kept in check as they would be in a natural ecosystem without the use of potentially harmful chemicals. Successful examples include the following:

   ➤ Lady bugs and lacewings released into a garden are voracious predators of mites, aphids, and many other common garden pests.

> Dragonflies are predators of mosquitoes. Because they are native to areas where mosquitoes dwell, simply attracting them to the area through environmental modifications (like a source of running water) may be all that is required.

iii. **Cons:** When it fails, biological control of pests can be a disaster. Before this method is tried, extensive research and trials should be done to ensure that the introduced species will not become a pest itself.

> Cane toads in Australia. The life cycle of the toad was not synchronized with the cane beetles it was meant to control. Cane toads are a generalist species, are highly toxic, and have become a threat to many native species. They continue to spread throughout Australia.

3. Other Natural Control Methods

i. Other natural control methods include using organic pheromone control (to disrupt mating), intercropping with plants like chrysanthemums known to repel pests, or using plant-derived sprays like neem oil as pesticides or insect repellents.

ii. **Pros:** When effective, natural control methods are an environmentally friendly alternative to more disruptive or toxic methods. Intercropping in particular has been effective in many home gardens and organic farms.

iii. **Cons:** Natural control methods may not be as effective or work as fast as inorganic methods. In many cases, use of natural controls requires a deep understanding of the complex natural relationships among species in order to implement them effectively. Timing or placement may be important to determining the outcome.

III. **Meat Production Methods**

A. Much of the meat protein human populations have traditionally consumed has come from open rangelands where the animals graze on grass. To meet the demands of a growing population, more meat is being raised on concentrated animal feeding operations (CAFOs).

B. The pros and cons of different methods of meat production include the following:

1. **CAFO Pros:** An inexpensive way to produce a large quantity of food quickly.

2. **CAFO Cons:** When raised at high density in less sanitary conditions, animals are more likely to become ill from contagious diseases and infections. To combat this problem, farmers often give their animals maintenance doses of antibiotics even when they are healthy. This is one factor that has contributed to the development of antibiotic-resistant strains of bacteria.

   i. To increase growth rates, animals may be fed steroid hormones. Many people are concerned that elevated levels of synthetic hormones in the meat they consume will interfere with their immune systems or disrupt their endocrine systems.

   ii. CAFOs or feedlots generate much organic waste, which can contaminate groundwater and surface water.

IV. Fisheries

A. Problems Related to Overfishing

   1. Many people in coastal regions of the world rely on ocean fisheries to supply protein in their diet.

   2. Overfishing has become a major problem in most areas where people catch and eat fish.

   3. Overfishing has led to the extreme scarcity of some fish species, which can lessen biodiversity in aquatic systems and harm people who depend on fishing for food and commerce.

B. Fishing Practices

   1. Many modern fishing practices have been developed to maximize the amount of fish that can be caught in a limited amount of time. Major drawbacks include bycatch (organisms such as sea turtles, sharks, and porpoises that are not the target but are killed anyway), environmental degradation, and pollution left when lines or nets break away from the boat and are set adrift. Here are some examples of modern fishing practices and their environmental drawbacks:

      i. Trawling. A trawling boat slowly drags a long, funnel-shaped, weighted net behind the boat along the ocean floor, essentially scooping up anything it encounters.

    ii.   Long lines. In this method, fishing lines with lengths of 50 miles or more are reeled out from the back of the boat. Large baited hooks are placed at intervals along the line to catch fish like swordfish, sharks, and tuna. The depth of the line can be adjusted to focus on different types of fish.

    iii.   Drift nets or gill nets. These types of nets are akin to installing a fence in the water to catch anything that swims through an area. These nets are often weighted on the bottom and attached to floating buoys on the top to stay upright.

C. Aquaculture

    1.   Aquaculture is the practice of raising a large concentration of fish in enclosed pens in coastal areas. Although it is very efficient, its high population density produces large amounts of waste that can contaminate water. Much like a CAFO or feedlot, diseases are common and can spread to wild populations.

Land and Water Use: Impacts of Resource Use

I. **Ecosystem Services of Forests**

A. Forests Provide a Host of Ecosystem Services

1. *Photosynthesis* is the process by which plants take in carbon dioxide (CO_2) and, using light from the sun as a source of energy, produce *glucose* (biologically available energy) and oxygen.

2. Providing commercially valuable products. Lumber from trees is used for building materials and fuel. Sap provides sweet syrups to eat and latex for glues and rubber. Other products from trees include nuts, such as pecans and cashews, and medicine and cellulose for use in a variety of products, from cosmetics to paper products.

3. Habitat for other species. Many species, from birds nesting in trees to shade-tolerant understory plants, depend on a closed-canopy forest for habitat.

   i. Forest ecosystems tend to have the highest biodiversity of all the biomes.

4. Atmospheric and microclimate effects. A closed-canopy forest has a distinct microclimate of shade and UV protection and higher humidity due to transpiration. In addition, the trees act as a windbreak.

   i. Large, mature forest trees also act as a sink for atmospheric carbon, not only by converting CO_2 during photosynthesis but also by accumulating it in their tissues, especially the trunks.

5. Improving soil and water quality. The roots of trees act as an anchor for soil, especially in areas with high precipitation. When trees are removed, water carries away topsoil, which in turn leads to erosion, mass movements (landslides), and desertification. By slowing the flow of water across the surface of the soil, trees also facilitate percolation of water through the layers of the soil,

increasing groundwater recharge and reducing pollution caused by runoff.

> *Ecosystem services is important and a great topic to bring up when answering a free-response question (FRQ).* **Ecosystem services** *are processes by which natural ecosystems or an environmental process produce resources that humans need. Examples include ocean currents moderating climate, insects pollinating crops, bacteria cycling nutrients through nitrogen fixation and decomposition, and all the examples described in this chapter. There is an increasing movement toward placing a dollar value on ecosystem services as a way to quantify the value of nature.*

B. Deforestation

1. *Deforestation* is the process of cutting down a significant portion of the trees in a forest and then converting that land to some nonforest use. Deforestation in tropical rainforests is a global problem. Forest land is cleared to grow crops, graze livestock, and use the wood for fuel.

   i. *Clear-cutting* is a type of logging in which all of the trees in an area are cut down regardless of size, age, or commercial value. This is the most damaging type of logging and, in most cases, leaves the land in such a condition that it is unlikely to regenerate naturally. Clear-cutting leads to fragmentation of the forest landscape and a decrease in biodiversity.

   ii. *Slash and burn* is an unsustainable farming method in which one area is clear cut and burned to enhance soil fertility for a short time.

2. *Sustainable Forestry* practices include reforestation, buying forest products produced through sustainable practices, and finding alternative materials to produce products made from harvested species.

C. Classifications of Forests Based on Levels of Disturbance

1. Old-growth forests, also known as primary-growth forests, have not suffered any major disturbances in hundreds or even thousands of years. They have very large, tall, old trees and show very few signs of human interference.

    i.   Old-growth forests have a closed canopy, a shady, open understory, and much biodiversity.

    ii.   Alaska has the largest area of old-growth forests in the U.S.

2. Secondary-growth forests are forests that have regrown after some major disturbance. Most U.S. forests are secondary-growth forests.

    i.   Typical disturbances include human activities (such as logging) and natural events such as storms, insect infestations, and fires.

    ii.   Forests are classified as secondary growth as long as there continues to be evidence of the disturbance, like dense understory growth and more light penetrating to the ground.

    iii. The term *jungle* is often used to describe secondary-growth forests.

D. Forest Fires

1. For most of U.S. history, forest management policy has been to extinguish any forest fire as quickly as possible. This is known as *fire suppression*. Fire suppression actually increases the risk of extremely hot and fast-burning wildfires because fallen limbs and underbrush accumulate and act as kindling once another fire starts.

2. Current forest management policy calls for periodic controlled burns, also known as *prescribed burns*. These small fires clear out any debris and brush that have accumulated. This reduces fuel buildup, returns nutrients to the soil, and in many fire-dependent ecosystems, actually stimulates the germination of seeds and maintains the conditions necessary for the ecosystem to thrive.

II. Impacts of Mining

A. Methods of Extraction

1. Depending on the resource being mined, and the depth and type of material around it, several mining techniques are used. They can be grouped into two categories: surface mining and underground mining.

    i.   *Surface mining* is the removal of large amounts of soil and rock, called overburden, to access the ore. These areas no longer provide wildlife habitat or recreational value and are prone to erosion.

> ➤ *Mountaintop removal* is a type of surface mining and is exactly what you would think—the top of a mountain is removed to expose the resource (usually coal) within the mountain.

B. Processing

1. Most of the materials extracted from the Earth are found as ores (rock containing a mixture of valuable minerals or metals and other nonvaluable materials) or impure mixtures in their natural state. Some processing is usually required to separate and purify the desired resource.

C. Impacts of Mining

1. The biggest environmental problem with resource extraction is waste disposal. Mining waste includes the soil and rock that were removed to access the ore. These tailings often contain sulfur which can lead to acid mine drainage and contamination of surface water and groundwater. Solutions to mitigate this problem include lining the tailing storage ponds and enforcing mine waste storage regulations.

D. Recycling vs. Mining

1. For most metals, recycling is the best choice. For example, the amount of energy needed to recycle an aluminum can into a new can is a small fraction of what is required to mine bauxite ore and refine it into usable aluminum. Aluminum, like most metals, can be recycled, with no loss in quality as a result of the process.

III. Human Impacts on Biodiversity

A. Habitat Fragmentation

1. A leading cause of loss of species is habitat fragmentation. This occurs when large tracks of continuous habitat are broken up into "habitat islands" often as a result of logging, road construction, and urbanization.

2. The impacts of habitat fragmentation can be mitigated by constructing habitat corridors. These provide landscape connections that link habitats that have been disrupted by human activities or structures. An overpass added over a newly built highway for animals to cross is one example.

IV. **Urbanization**

A. Impacts of Urbanization

1. When natural land surfaces are replaced with impervious surfaces, water can run off and cause flooding instead of infiltrating and recharging groundwater.

B. Urban Microclimates/Heat Islands

1. More asphalt and concrete changes the albedo (reflectivity) of ground surfaces, causing more heat to be absorbed during the day. The heat then radiates back into the atmosphere during the night, keeping nighttime temperatures in cities warmer than in the surrounding natural areas.

2. Urban areas have fewer plants to provide shade and the possible cooling effects of transpiration.

3. More people using air conditioning, cars, and machinery means more heat is generated. Heat is a by-product of combustion.

C. Urban sprawl occurs when population spreads from high-density cities to lower-density urban areas.

D. Sustainable Construction Practices

1. Because urban areas are usually growing, many of our most creative construction alternatives arise in cities.

   i. LEED certification. The U.S. Green Building Council offers a LEED (Leadership in Energy and Environmental Design) certification program to guide people in the construction of all types of buildings. The LEED program covers energy and water efficiency, sustainable building materials, reductions in indoor pollution, and environmental impact of the building site.

   ii. Cogeneration is the process of recapturing waste heat generated by industrial processes for further uses. For example, waste heat generated by power plants can be redirected to heating homes.

REFERENCES

National Park Service (NPS)

NPS Units by Type. *http://www.nps.gov/legacy/nomenclature.html*

NPS Overview. *http://www.nps.gov/news/loader.cfm?csModule=security/ getfile&PageID=387483*

Bureau of Reclamation: About Us. *http://www.usbr.gov/main/about/fact. html*

U.S. Forest Service: About Us. *http://www.fs.fed.us/aboutus/*

National Wilderness Preservation System of the United States. *http:// www.nationalatlas.gov/mld/wildrnp.html*

Wetlands Definitions/wetlands/U.S. EPA. *http://water.epa.gov/lawsregs/ guidance/wetlands/definitions.cfm*

EPA Wetlands overview *http://water.epa.gov/type/wetlands/outreach/ upload/overview.pdf*

USGBC Intro—What LEED Is. *http://www.usgbc.org/DisplayPage. aspx?CMSPageID=1988*

Energy Resources and Consumption

Chapter

12

I. Renewable and Nonrenewable Energy Sources

A. Renewable energy resources are those that can be replenished at the rate they are consumed. Solar and wind energy are examples of renewable resources.

   1. Nonrenewable energy resources are those that are present in a fixed supply on Earth and cannot be easily replaced. Fossil fuels are nonrenewable.

B. Energy Types

   1. Energy exists in two major forms, potential and kinetic.

      i. *Potential energy* is stored energy that can be converted to active or kinetic energy given the right situation.

      ii. The total of all the energy of an object based on its movement (kinetic) or position (potential) is *mechanical energy*.

   2. Potential Energy

      i. Energy can be stored for later use in several ways. For AP® purposes, the three most significant types of potential energy are gravitational, chemical, and nuclear.

         ➤ *Gravitational potential energy* is the energy of an object based on its position and the force of gravity acting on that object. For example, rivers tend to flow from higher elevations to lower elevations. Water is pulled downstream by the force of gravity. If a dam is built to block the flow of a river, the water accumulating behind the dam now has potential energy. As long as the dam holds the water back, the water continues to accumulate gravitational potential energy. If the dam is opened and the water is allowed to flow, the potential energy of the accumulated water is converted to kinetic energy as the water rushes down the river.

➤ *Chemical potential energy.* Much of the energy contained within biological systems, like our bodies, is stored in the chemical bonds of molecules like fats and glucose within our cells. When our cells need to unlock that potential energy, they initiate the process of cellular respiration to convert those molecules into adenosine triphosphate (ATP), which concentrates this energy into larger, more useful packets. As high-energy phosphate groups are removed, from the ATP molecule—to make Adenosine Diphosphate (ADP), then Adenosine Monophosphate (AMP)—energy is released to be used as kinetic energy to power the body's metabolism.

➤ *Nuclear potential energy* is contained within the nucleus of an atom. Some atoms spontaneously split apart in a process called *nuclear fission*. Others can be compelled to nuclear fission through collision with a rapidly moving neutron. When this happens, a great deal of energy is converted from potential to kinetic energy. This process is useful in nuclear reactors for generating electricity.

3. Kinetic Energy

   i. *Kinetic energy* is the energy of motion.

   ii. All moving objects have kinetic energy.

   iii. When the potential energy of matter is put into motion as kinetic energy, work can be done.

C. Forms of Energy

   1. *Thermal energy (heat).* As matter increases in temperature, the molecules of the atoms making up that matter gain kinetic energy and tend to spread farther apart. This process may result in changes of phase from solid to liquid, to gas, to plasma.

2. *Electrical energy* is negatively charged electrons traveling from one place to another through a conductor. Electrical energy is useful for moving energy across a distance (for example, from a power plant to your home). Once the energy arrives at the desired destination, it can be converted to heat, light, or mechanical energy.

3. *Chemical energy* is stored in the bonds between atoms. For example, the energy locked in the carbon–hydrogen bonds of fossil fuels can be converted to (thermal) kinetic energy by burning coal or oil to break those bonds.

D. Units of Measuring Energy

1. The *British thermal unit* (BTU) is the amount of heat required to raise one pound of water by one degree Fahrenheit. (This is the most common unit of energy used on the AP® exam.)

2. The *joule* is the energy needed to lift an object weighing one Newton by one meter of distance. (The Newton is the metric unit of weight.)

3. A *calorie* is the amount of heat needed to raise the temperature of one gram of water by one degree Celsius. This unit is often confused with food calories (often written as Calorie and called "big calories"), which are actually kilocalories (1,000 calories per 1 Calorie).

4. A *kilowatt hour* (kWh) is the amount of energy used to produce one kilowatt of electricity for a continuous hour. This is the unit most commonly seen on home electricity bills.

Test Tip

You need not memorize anything about these units! You may see an energy conversion problem on the exam, but any conversion factors you need will be provided. The units are covered here so that you can become familiar with the terms.

Math Practice

Converting energy units is easy as long as you remember to set up your equation so the units you have cancel out, leaving you with only the desired unit. Here are a few examples for practice.

1. *Suppose a 1,450-watt microwave is used for 30 minutes each day for one year. How many kWH per year does the microwave use?*

2. *1 kWH of electricity is equal to 3,413 BTU. One cubic foot of natural gas supplies approximately 1,031 BTU of energy. How much natural gas would be needed to power the 1,450-watt microwave for one year?*

3. *1 therm is equivalent to 100,000 BTU. How many therms of natural gas are needed to power the 1,450-watt microwave for one year?*

ANSWERS:

1. $1450 \text{ watts} \times \dfrac{1 \text{ kW}}{1000 \text{ watts}} \times \dfrac{0.5 \text{ hours}}{\text{day}} \times \dfrac{365 \text{ day}}{\text{year}} = \dfrac{265 \text{ kWh}}{\text{year}}$

2. $\dfrac{265 \text{ kWh}}{\text{yr}} \times \dfrac{3413 \text{ BTU}}{1 \text{ kWh}} \times \dfrac{1 \text{ ft}^3 \text{ natural gas}}{1031 \text{ BTU}} = \dfrac{877 \text{ ft}^3 \text{ natural gas}}{\text{yr}}$

3. $\dfrac{265 \text{ kWh}}{\text{yr}} \times \dfrac{3413 \text{ BTU}}{\text{kWh}} \times \dfrac{1 \text{ therm}}{100,000 \text{ BTU}} = \dfrac{9.04 \text{ therm}}{\text{yr}}$

E. Thermodynamics

1. *Thermodynamics* is the study of how energy flows through natural systems.

2. First Law of Thermodynamics. The law of conservation of energy states that the amount of energy in the universe is constant. Energy can neither be created nor destroyed. Energy may be transferred from one form to another, but it is never lost and cannot be created anew.

3. Second Law of Thermodynamics. The law of entropy states that, as energy changes from one form to another, it is always moving from a state of more organization to one of less organization.

i. As energy becomes less organized and concentrated, its use-fulness for work is also decreased. (Remember trophic pyramids?) This law is the reason why there are usually no more than five trophic levels in any food web.

ii. This law explains why a perpetual motion machine is impossible. As a machine does work, some of the energy in the system is always lost to friction or heat, so eventually more energy will have to be added from outside the system to allow the machine to continue working.

iii. This law also explains why energy always flows from an object of higher temperature to one of lower temperature.

II. Energy Consumption, Conservation, and Efficiency

A. Trends in Global Energy Consumption

1. Earth processes that determine the formation and location of natural energy resources means the distribution of ores and fossil fuels is not uniform.

2. There has been a global energy transition from wood-coal-oil-natural gas and increasing use of renewables such as wind and solar.

3. Energy resource use differs from developed to developing countries; fossil fuels are the most widely used source of energy.

4. As countries become more industrialized, their demand for fossil fuels increases.

5. Energy use decisions are based on availability, price, and government regulations and policies.

B. Strategies for Conserving Energy

1. Reducing waste in our use/choices. By examining your energy use habits, you will probably find that you are wasting energy all the time. By raising your awareness of conservation, you can greatly reduce the energy used without affecting anything you need.

i. Heating and Cooling Your Home

➤ Install a programmable thermostat to use less air conditioning during the hours of the day when no one is home.

➤ Seal leaky windows, doors, and windows with weather stripping or caulk to avoid air leaks that cause the air conditioning to work more. Add insulation to walls and ceilings and use windows that are double-paned or low-emissive to avoid heat/cooling losses.

➤ Pay attention to the exterior color of your home and roofing material. In warm climates, light colors reflect the heat, helping the house stay cooler. In cold climates, darker colors absorb more heat, helping the house stay warm.

ii. Appliances

➤ Even when turned off, many appliances still use electricity as long as they are plugged in. Large appliances like computers, televisions, and stereo systems should be plugged into a power strip that can be turned off when the appliances are not in use.

➤ Even in sleep mode, computers and televisions use energy. Turn off these appliances at night and when you will not be using them for several hours. If you use a computer at school or work, turn it off (including the screen) before you leave for the day.

➤ Turn off lights and other appliances when you are not in the room. Do a quick walk through your house before you leave for the day to ensure that ceiling fans, radios, and lights are not left on.

2. Increase the efficiency of existing technology. In the last few years, the energy efficiency of lights and appliances has been greatly improved. Before buying new products, compare the energy efficiency of your options. Consider not only the purchase price, but also the long-term cost of using that product every day.

i. Compact Fluorescent Lights (CFLs) and light-emitting diodes (LEDs). When choosing a replacement light bulb, many people still buy incandescent bulbs out of habit. They cost less to purchase (less than a dollar compared to $3 or $5 per CFL). Some people prefer the quality of light produced by incandescent lights. The reality is that CFLs and LEDs are far and away the best lighting option.

    ➤ Only 10 percent of the energy used by incandescent bulbs produces light. Ninety percent of the energy used by the bulb generates heat! That is why you should never touch an incandescent bulb that has been turned on.

    ➤ Recently improved CFLs last, on average, ten times longer per bulb and use 75 percent less energy than incandescent bulbs.

    ➤ LEDs (light-emitting diode) have many advantages over other light bulbs including lower energy consumption, longer lifetime, and unlike CFLs, do not contain mercury.

ii. Energy Star is a rating system that allows consumers to compare the energy use of new appliances to one another. Energy Star-certified appliances meet minimum energy conservation criteria and are good choices when considering the long-term cost of a new appliance.

iii. Gas mileage in cars. Improvements in lightweight body materials, more efficient engines, and better design, mean many cars are getting more than 30 miles per gallon on the highway. Options like hybrids and electric cars offer even greater energy efficiency.

Math Practice

Efficiency problems are simply a matter of working with proportions and percentages. Here are a few practice questions.

1. *If your home uses twenty 100-watt incandescent light bulbs for four hours per day, how many kilowatt-hours of electricity are needed to power the bulbs for one year of use?*

2. *Compact fluorescent light bulbs (CFLs) are 75 percent more efficient than incandescent bulbs. How many kilowatt-hours per year can you save by replacing 20 incandescent bulbs with CFLs?*

3. *A compact hybrid gets 51 miles per gallon of gasoline in highway driving. A luxury sedan averages only 25 miles per gallon on the highway. If gas costs $3.00 per gallon, how much money will a person who drives 10,000 miles per year save by choosing to purchase the hybrid?*

ANSWERS:

1. Your answer needs to be in units of kilowatt-hours per year. As you set up your fractions, make sure the units line up to cancel each other, leaving your desired units in the right places!

$$20 \text{ bulbs} \times \frac{100 \text{ watts}}{\text{bulb}} \times \frac{1 \text{ kW}}{1000 \text{ watts}} \times \frac{4 \text{ hours}}{\text{day}} \times \frac{365 \text{ days}}{\text{year}} = \frac{2920 \text{ kWh}}{\text{year}}$$

2. Seventy-five percent more efficient means that only 25 percent of the energy needed to run incandescent bulbs is needed to run CFLs. Simply multiply your answer to Math Practice 1 by 25 percent.

$$\frac{2920 \text{ kWh}}{\text{year}} \times 0.25 = \frac{730 \text{ kWh}}{\text{year}}$$

3. First, note that your answer needs to be in the units of dollars per year. Set up your fractions to cancel units, making sure dollars is on the top of the fraction and year is on the bottom. The rest falls into place.

$$\text{Compact hybrid} = \frac{10,000 \text{ miles}}{\text{year}} \times \frac{1 \text{ gallon}}{51 \text{ miles}} \times \frac{\$3}{\text{gallon}} = \frac{\$588}{\text{year}}$$

$$\text{Luxury sedan} = \frac{10,000 \text{ miles}}{\text{year}} \times \frac{1 \text{ gallon}}{25 \text{ miles}} \times \frac{\$3}{\text{gallon}} = \frac{\$1200}{\text{year}}$$

$$\text{Savings} = \text{cost for sedan} - \text{cost for hybrid} = \frac{\$1200}{\text{year}} - \frac{\$588}{\text{year}} = \frac{\$612}{\text{year}}$$

III. Fossil Fuels

A. Fossil fuels—coal, oil, and natural gas—provide the most commercial energy globally.

    1. The combustion of fossil fuels releases CO_2, a greenhouse gas.

    2. When fossil fuels are burned, the energy is used to heat water, which generates steam, which turns a turbine to generate electricity.

    3. Hydraulic fracturing (fracking) is used to release natural gas (methane-CH_4) from shale rock. Fracking also produces crude oil. Although methane is a cleaner-burning fuel than coal and oil, the process of fracking has the potential to contaminate groundwater.

B. Coal

    1. The primary use of coal is to generate electricity in large, coal-fired power plants. Coal can also be used in individual homes for heating.

    2. Formation of coal. Over time, heat and pressure increase and the moisture content of compressed plant material decreases, a process that leads to the eventual formation of coal. Classes of coal (listed in order of age and purity) include peat, lignite, bituminous coal, and anthracite.

        i. Peat is the first step in coal formation and is not technically coal. Carbon-rich soils concentrate in bogs and fens and may be harvested to burn for fuel.

        ii. Brown and soft, lignite is a form of coal with a low carbon content.

        iii. Bituminous coal is commonly used to generate electricity, or it is converted into coke, which is necessary for steel making.

        iv. Anthracite is the purest form of coal, up to 97 percent carbon. This is the cleanest-burning coal due to low levels of impurities like sulfur and nitrogen.

    3. Extraction. Coal is commonly mined in open-pit mines or strip mines (often resulting in mountaintop removal), although there are also underground coal-mining operations.

4. Pollution. To release its energy, coal is burned at high temperatures in order to heat water. The heated water evaporates into steam, which is used to turn turbines connected to electromagnets, which then produce an electric current. Burning coal is a major source of air pollution in MEDCs. Here is a list of some of the pollutants released into the atmosphere as a result of burning coal to produce electricity.

   i.   Carbon dioxide (CO_2). Coal-fired power plants are the major source of anthropogenic carbon emissions. The CO_2 released by coal-fired power plants is a major contributor to global climate change, far more than car exhaust systems or deforestation.

   ii.   Particulate matter. Small particles of partially combusted ash are released from power plants. Particulate-matter pollution leads to respiratory problems, like asthma and chronic bronchitis, and contributes to smog formation.

   iii.   Sulfur oxides (SO_x) and nitrogen oxides (NO_x). SO_x and NO_x are also by-products of combustion. In the atmosphere, they combine to form sulfuric and nitric acids, which lead to damaging acid deposition.

   iv.   Heavy metals, arsenic, lead, cadmium, and mercury, are either released into the atmosphere or concentrated into waste ash. If waste ash is not properly treated and stored, these chemicals can leach into the environment, causing contamination of local freshwater (including groundwater).

5. Wet scrubbers are installed in smokestacks to reduce emissions of sulfur. Electrostatic generators use static electricity to capture particulate matter. Selective catalytic reduction units are used to reduce emissions of NO_x.

6. Coal reserves. More than 50 percent of the world's coal reserves are located in China, the U.S., and Russia (including the former Soviet republics). It is estimated that at current consumption rates, the world's coal reserves will be depleted within 150 years.

C. Oil/Petroleum

1. In its natural state, crude oil is a thick, sticky liquid that is formed from the remains of oceanic organisms that lived millions of years ago. Over time, these remains were heated and compressed to form oil. Once it is extracted, crude oil is refined to make a variety of different types of fuels and products.

2. Crude oil is extracted from tar sands found in the boreal forest of northern Canada. This process of recovering oil has a high environmental impact and poses a health risk to local residents who are exposed to heavy metal contamination in their food.

D. Natural Gas

1. Like coal and oil, natural gas is formed from the heated and compressed remains of ancient organisms.

   i. Natural gas deposits are often discovered with deeper oil deposits, and when natural gas is found alone, it is often more than 2 miles below the Earth's surface.

   ii. Natural gas is composed primarily of methane, although it also contains propane, ethane, and butane in smaller amounts.

2. Uses of natural gas. Because it is relatively free of impurities, natural gas is the cleanest-burning fossil fuel.

   i. It is commonly used in home applications like heating, cooking, and powering appliances.

   ii. Although burning natural gas releases fewer pollutants like SO_x, NO_x, and carbon dioxide, natural gas leaks are composed of almost pure methane. *Pure methane* is an aggressive greenhouse gas in its own right—about twenty times greater than carbon dioxide!

IV. Nuclear Energy

A. Nuclear energy is released by breaking apart the nucleus of an atom.

1. The most common element used to generate nuclear energy is uranium, specifically an isotope called U-235. Although uranium is a nonrenewable mineral resource, supplies are plentiful.

2. When a neutron particle is fired at high velocity at the nucleus of a U-235 atom, the collision causes the nucleus to break apart, releasing a great deal of energy.

3. Nuclear energy is considered a clean energy resource because no CO_2 is produced from nuclear reactions, unlike combustion reactions.

B. Nuclear power plants generate electricity by using the energy released from the fission of U-235 to heat water, create steam, spin a turbine, and generate electricity.

C. Problems with Nuclear Power

1. Uranium-235 has a long half-life and will remain radioactive for a long time. The greatest risk to environments and human health is associated with how to safely store the nuclear waste.

2. Half-life. All radioactive elements have a unique half-life, the time it takes for half of the atoms in a sample to decay to a more stable form. For example, the carbon-14 isotope has a half-life of about 5,700 years. For a sample of 100 atoms of C-14, after 5,700 years, there will be 50 atoms of C-14 remaining in the sample. After another 5,700 years there will be 25 C-14 atoms left; after another 5,700 years, approximately 12.5 atoms remain. This goes on infinitely, with half of the remaining atoms decaying at each passage of the half-life. We measure remaining radioactive atoms or radiation emitted by these atoms when we look at half-life calculations.

You may see a half-life calculation in the multiple-choice section. Don't make the common mistake of thinking that, if 50 percent decays after one half-life, the remaining 50 percent will be decayed after the second passage of half-life. It is an exponential decay, and half of what remains is lost at each step.

D. Nuclear Power Plant Accidents

1. Three Mile Island, Pennsylvania (1979)

    i. The Three Mile Island (TMI) nuclear power plant experienced a series of failures in the mechanical operation, design, and human communication in the plant, which led to a partial meltdown of the reactor core.

2. Chernobyl, Russia (1986)

    i. Due to a design flaw and operator error, the Chernobyl nuclear power plant experienced a total meltdown of its core and released a significant amount of radioactive gasses into the environment.

3. Fukushima Daiichi Nuclear Power Plant, Japan (2011)

   i. In the wake of a magnitude 9.0 earthquake and subsequent tsunami, three reactors at the Daiichi power plant overheated, causing an explosion that released radioactive particles into the surrounding area.

E. Pollution

   1. Zero emissions. Unlike fossil fuels, nuclear power does not depend on combustion (burning) to generate energy. As a result, it produces no emissions of greenhouse gases in its operation.

   2. Radioactive waste. All power plants generate radioactive wastes, which can be harmful for thousands of years. Because of this, proper storage of this waste must be designed to withstand great passages of time and minimal human intervention to ensure its safety.

   3. Thermal pollution. Water is essential to nuclear power plants because it is used to circulate around the reactor as a cooling agent. As a result, nuclear power plants are generally located near large bodies of water that may be drawn into the plant and then released back into the environment after circulating through the plant. If water is released directly to the environment without first being allowed to cool down, the high-temperature water may be harmful to organisms living in the body of water.

Math Practice

After 80 million years have passed, only $\frac{1}{128}$ of the original number of radioactive atoms of an element in a sample remains. What is the half-life of the element?

ANSWER:
Start with the proportion $\frac{1}{128}$. This is a passage of 7 half-lives of the element ($\frac{1}{2}, \frac{1}{4}, \frac{1}{8}, \frac{1}{16}, \frac{1}{32}, \frac{1}{64}, \frac{1}{128}$).
Next, divide 80 million years by 7 to get your answer: 11.4 million years.

V. Renewable Energy Resources

A. Renewable resources can be regenerated (such as trees and forests) or recycled infinitely (water). A perpetual resource such as the sun's rays, is one that cannot be exhausted by human activities regardless of how much is used.

B. Solar Energy

1. Ultimate source for all other types of renewable and many nonrenewable energy sources. It can be used to heat buildings or generate electricity.

2. Heat from the sun drives the water cycle by causing evaporation and transpiration, thus making hydroelectric power possible.

3. Differences in air temperature and pressure caused by uneven heating of the Earth's surface generate winds.

4. Light from the sun makes photosynthesis possible, which in turn makes it possible for biomass to accumulate.

5. Even fossil fuels were ultimately made possible from the sun's energy.

C. Active Solar Energy

1. *Photovoltaic cells* are used to collect and convert energy from the sun directly into electricity.

2. Active solar heating involves heating a liquid; the heat is then stored and distributed through heat pumps and fans to heat a building.

3. Large-scale solar thermal uses mirrors to capture energy, which is used to heat water, make steam, spin a turbine, and generate electricity.

D. Passive Solar Energy

1. By using the sun's energy directly to accomplish a task, passive solar heating allows for a highly efficient source of energy.

   i. No mechanical, photovoltaic, or chemical conversions are used in passive solar energy.

   ii. Passive solar energy is most commonly used for heating systems.

2. Building design. A greenhouse is a good example of passive solar heating. The clear panes of glass allow the sun's rays to penetrate and warm the air inside, but the longer wavelengths of the infrared radiation (heat) cause it to be trapped inside, warming the greenhouse far beyond outside temperatures.

   i.   Buildings are designed with windows facing the east, west, and south, so the sun's capacity for heating them can be maximized.

   ii. Using dark colors in areas where heat is desired increases the effectiveness of heat absorption. For example, dark-colored shingles can reduce heating bills in northern climates.

   iii. Conversely, planting shade trees, using reflective window coverings, and using light colors for roofing materials can greatly reduce cooling expenses in warmer climates by reducing a building's heat absorption potential.

3. Solar thermal systems can be used to generate electricity by heating water to make steam that turns turbines. In these systems, a series of mirrors are used to concentrate sunlight in order to heat the water efficiently.

E. Benefits of Using Solar Energy

1. Zero emissions and pollutants. Because nothing is burned (at least, not on Earth!) to produce electricity, all of the combustion-related pollutants are avoided by using solar power.

2. Habitat preservation. No mining or transport of fuel is required. As a result, harmful practices such as mountaintop removal and pit mining are avoided. Because many solar applications can be installed directly on rooftops in already developed areas, animal habitats are not disturbed.

3. Efficiency. Passive solar heating is extremely efficient because of the simplicity of energy conversions. It only requires a low initial monetary investment, especially when passive solar design elements are incorporated into the plans of a new construction project.

4. Batteries store excess energy for use during times when the sun is not out. Manufacture and disposal of these batteries is another environmental concern, especially because modern long-life batteries use heavy metals like cadmium, lead, and mercury. All of these heavy metals are considered powerful toxins when they are released into the environment by improper disposal.

VI. Hydroelectric Power

A. Benefits of Building Dams

1. For centuries, dams have been built to redirect water to do work. Dams allow us to direct water toward our crops for irrigation, control the annual flooding cycle of large rivers, and produce electricity.

2. Energy generation

   i. Powerful rivers, such as the Colorado River in the western U.S., have been prime sites to build hydroelectric dams. Water is stored in the reservoir and moving water spins enormous turbines to generate electricity.

   ii. Turbines can also be placed in rivers where the moving water spins a turbine.

   iii. Tidal power harnesses the energy produced by changing tides to spin a turbine and generate electricity.

3. Benefits of building a dam

   i. Rivers with no dams often have a seasonal fluctuation of water. In the spring, torrents of water are released as snow and ice melts in the mountain source areas. In addition, spring rainfalls often lead to flooding. In winter, the weather is drier, and most precipitation falls as snow and ice in the mountains, so the flow of the river is greatly decreased.

   ii. Capturing some river water in a reservoir behind the dam creates a year-round source of water, which can be used to irrigate crops and supply water to homes and businesses.

   iii. The reservoirs that accumulate water behind large dams create large lakes used for recreation.

B. Negative Effects of Dams

1. Water diversion. Aside from producing electricity, dams are also useful for changing the route of the flow of water, known as *water diversion*. But this also removes water needed for a healthy, instream ecosystem.

2. Reduced Fertility of Floodplains

   i. The annual cycle of water flow. While inconvenient to humans, the natural cycle of flooding and drought common to large rivers is an important part of maintaining soil fertility in land adjacent to the river.

ii. Disruption of aquatic organisms. In addition to changing the amount of water that flows in a river, dams also create a physical barrier to organisms attempting to navigate along the river.

➤ Salmon are a commercially important migratory fish that cannot complete their life cycle and breed when they are blocked by a dam. Some dams incorporate "fish ladders," a type of stream flowing over the top or around the side of the dam to facilitate migration of fish and other aquatic life.

VII. Wind Power

A. To generate electricity from wind power, the kinetic energy from the moving air turns a turbine; mechanical energy is converted into electricity.

B. Benefits of Wind Power

1. No emissions. Again, because nothing is being burned, no atmospheric emissions are released in the energy-generating process.

C. Drawbacks of Wind Power

1. Manufacturing the turbines. Most modern wind turbines are made of metals and plastics, so there are negative environmental effects associated with extracting and manufacturing the materials needed for their construction.

2. Spinning blades. Because energy is generated by large spinning blades, organisms like birds and bats can be killed by contact with the blades. This can be mitigated by tracking migratory routes and siting wind turbines where they will have minimal impact on wildlife.

VIII. Biomass

A. Burning fuel wood or charcoal biomass is common in developing countries. It relies on unsustainable logging practices and produces air pollutants such as CO and CO_2. Its particles harm those who use wood or charcoal for cooking.

B. Crops like corn and sugar cane can be harvested and fermented into ethanol that can be burned directly or added to gasoline. Although it still produces CO_2 during combustion, ethanol does not add additional carbon to the atmosphere.

IX. Geothermal Energy/Hydrogen Fuel

A. *Geothermal energy* is energy from below the Earth's surface.

B. Geology Review

1. According to the theory of plate tectonics, the interior of the Earth is divided into layers of rocks and minerals that vary by temperature and density, basically becoming warmer with increased proximity to the innermost core. Just below the surface crust are large deposits of superheated rock called *magma* formed primarily as a result of the radioactive decay of atoms in the Earth's core.

2. The Earth's crust is made up of large tectonic plates that slowly move across the asthenosphere. They are driven by convection currents within the Earth's mantle. At the boundaries where these plates interact, the crust tends to be thinner, allowing this heat to rise closer to the surface of the crust, often resulting in volcanoes.

C. Conversion to Useful Energy

1. The simplest way to capture geothermal energy is to dig a pipeline down into the crust to a hot spot and send water down to be superheated. The water is then returned to the surface as steam to power a turbine. As the steam cools and condenses back to water, it is returned underground to continue the cycle.

X. Energy Conservation

A. Electric and Hybrid Cars

1. All gasoline-powered cars have a battery. The internal combustion engine burns fuel to power the drivetrain of the car. This keeps the battery charged to run all of the car's electrical components.

2. In an electric car, the battery does all the work. The battery is charged between uses by plugging it into an electrical power source.

3. Hybrid cars work on the same principle, but they also have a smaller internal combustion engine to use as a backup to extend the range of the battery.

4. At present, battery technology cannot produce a battery with a long enough life to allow for long car trips without frequent recharging.

B. Hydrogen Fuel Cells. When hydrogen and oxygen from air react to form water, energy is released as well. Although the product, water vapor, is a greenhouse gas, it is short-lived in the atmosphere. Thus, hydrogen fuels cells are seen as a good alternative to cars powered by gasoline.

When you assess the relative merits of each type of energy, a few general rules hold true:

1. There is no perfect source of energy. The best way to protect the environment is to use less energy, period. Alternatives such as solar power and wind power require harmful mining and extraction processes to obtain the raw materials needed to make the necessary equipment or machinery.

2. Regardless of origin, any fuel that must be burned will produce atmospheric pollution.

3. Renewable energy does not automatically equal environmentally-friendly energy. For instance, some biomass fuels produce far more pollution and negative environmental effects than some nonrenewable sources.

REFERENCES

Light bulbs (CFL): Energy Star. *http://www.energystar.gov/index.cfm?fuseaction=find_a_product.showProductGroup&pgw_code=LB*

Atmospheric Pollution: Environmental Toxicology and Pollution

I. Toxicity

A. Measuring Toxicity

1. One measure of the toxicity of a substance is to test the effects of exposure on nonhuman organisms. Whether or not you find animal testing ethical, every medicine approved by the U.S. Food and Drug Administration (FDA) must first be tested on animals to determine its possible harmful side effects.

2. LD50 is the amount of a substance that will be a lethal dose to 50 percent of a test population. Usually measured in mg of substance per kg of body weight, the lower a substance's LD50 value, the greater its toxicity (meaning a smaller dose will be fatal). Because LD50 is measured in mg per kg of body weight, we can extrapolate values obtained on small mammals like rodents to estimate a range of lethal doses for humans. Note, however, this measurement is not always accurate because of physiological differences between different species.

3. Dose-Response Curve

   i. A dose-response curve measures the mortality of a toxin or drug based on the dose or concentration.

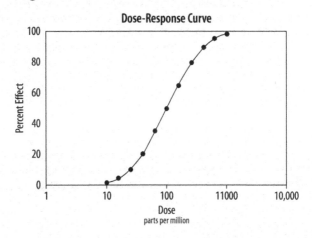

Dose-Response Curve

Percent Effect

Dose
parts per million

B. Synergy

1. In some cases, two or more seemingly harmless substances have a toxic effect in combination. Another way to describe this is "the whole is greater than the sum of its parts."

2. Synergistic toxicity is of particular concern in aquatic ecosystems where various pesticides, heavy metals, and other toxins accumulate from industrial, roadway, and agricultural runoff. Although government agencies have set toxicity limits for many individual chemicals, the synergistic effects of many of these in combination are unknown.

C. Threshold of Toxicity

1. The *threshold of toxicity* is the limit below which no harmful effect is seen. Once the toxic threshold is reached or exceeded, harmful effects begin to appear.

D. Bioaccumulation and Biomagnification

1. *Bioaccumulation* is the process in which an organism absorbs a toxin, usually fat soluble, in high concentrations.

   ➤ For example, bracken ferns are fast-growing, hardy plants that, when planted in soils contaminated with toxins such as arsenic, can absorb the contaminant through their roots and store it in their leaves and other tissues. When harvested and tested, the ferns have levels of arsenic in its tissues that far exceed the levels in the surrounding soils.

2. *Biomagnification* is the process where an accumulated toxin is passed up the food chain as one organism eats another.

   ➤ For example, DDT is a pesticide that is very effective at killing mosquitoes. In low doses, DDT is not particularly harmful to larger organisms, so it was widely used throughout the tropics, the U.S., and Europe. In the 1960s, scientists noticed that large predatory birds were declining in record numbers because there was something interfering with the calcium in their eggs, leaving them soft and prone to breaking. When affected birds were tested, alarmingly high levels of DDT were present in their tissues.

   ➤ Here's how biomagnification in the case of DDT works:

      — DDT was sprayed in the water to kill mosquito larvae. Small zooplankton and phytoplankton species accumulated small doses of DDT in their cells (one dose each, for example).

— Small fish eat plankton. Each time a fish eats a plankton, it acquires the plankton's one dose of DDT in its own body, in addition to any DDT it would take in merely by being exposed to the sprayed water. Let's say that each small fish has 10 additional doses of DDT from eating plankton, plus 1 dose from living in the sprayed water, equaling 11 doses of DDT per fish.

— If a large fish eats 20 small fish, it also gets the DDT accumulated in the tissues of each small fish it eats. So, it actually ingests 220 doses in addition to the 1 dose it receives from living in sprayed water, equaling 221 doses of DDT per large fish.

— Along comes an eagle that catches and eats the big fish. For every fish eaten, the eagle gets a dose of 221 units of DDT. If she eats only 5 fish, that's a total of 1,105 units of DDT. Keep in mind that birds have a relatively low body weight for their size; it is no wonder they were being harmed by the small amounts of DDT in the environment.

➤ Now you see how even a low concentration of a toxin can wreak havoc on top predators in a food web due to biomagnification.

II. Pollution

Pollution is the addition of anything into a natural environment that may cause harm. Pollution is related to toxicity because the more toxic a substance, the greater the potential damage to the environment. Aside from toxicity, a few other factors help to determine how much damage a substance might do.

A. Persistence

Persistence means how long a substance remains in an environment.

1. Measuring the half-life of a radioactive isotope, such as uranium 238, is an example of measuring persistence.

2. Another way of measuring persistence is to look at how long it takes for a substance to disintegrate. For example, if you toss a banana peel on the ground and come back several weeks later, the peel will likely be completely gone. Organic materials that can be readily decomposed by bacteria have a very low persistence in the environment.

3. If you drop a plastic bottle instead of a banana peel, that bottle will be relatively unchanged. You could conceivably come back years later and find it looking very much the same—dirty perhaps, but a plastic bottle all the same. Synthetic materials have a very long persistence in the environment.

III. Point vs. Nonpoint Sources of Pollution

A. As discussed in Chapter 3, point sources of pollution are easy to identify. Literally, you can point to the smokestack of a factory and say, "That's where the pollution is originating."

1. Effluent pipes carrying industrial waste from factories into waterways used to be common point sources.

2. The oil rig Deepwater Horizon's explosion and oil leak in the Gulf of Mexico in 2010 was a dramatic point source of oil pollution.

B. Nonpoint sources of pollution are more difficult to identify.

1. Runoff is a major nonpoint source pollutant. Agricultural runoff is water that accumulates nutrients such as nitrates and phosphates from fertilizers or animal wastes, as well as toxic chemicals found in herbicides and pesticides. These chemicals dissolve in the water and are carried to waterways.

2. Acid deposition is another nonpoint source pollutant. Nitric and sulfuric acids are formed in the atmosphere from NO_x and SO_x (respectively) emitted from power plants and industrial operations. The acid rain or other types of deposition that return to the ground and water may be miles away from the original emissions, and they may contain a mixture of acids formed from different airborne pollutants.

3. One characteristic common to all nonpoint source pollutants is that they rarely contain toxins from a single event or location. They usually contain a mix of contaminants from several sites that mix together in the environment.

It is important to see how many of the issues we study in Advanced Placement® Environmental Science are interrelated. For example, our choices of which energy sources to use will have a huge impact on the levels of pollutants. Sometimes the toxicity of potential pollution drives the choices we make. The Three Mile Island nuclear disaster totally changed the public outlook concerning the desirability of nuclear energy.

IV. Pollutants Across Air, Water, and Land

A. Heavy Metals

1. Heavy metals such as lead, mercury, cadmium, and cobalt are potent neurotoxins, and low-level exposures may be harmful. As technology advances, we are encountering more heavy metals in our computers, cell phones, and batteries.

2. Improper disposal of items containing heavy metals can lead to leaching into the soil and groundwater.

3. Heavy metal pollution is even an atmospheric problem. Burning coal releases mercury that can be transported and deposited large distances from the source of emission. When elemental sources of mercury enter aquatic environments, bacteria in the water convert it to highly toxic methylmercury.

B. Radioactive Wastes

1. Radioactive isotopes emit potentially harmful alpha, beta, and gamma waves that can enter cells and disrupt DNA, making exposure very dangerous. These waves move through the air, so direct contact is not necessary to cause injury.

2. Soil contaminated with ash from a radioactive exposure can cause growing food plants to bioaccumulate harmful radiation, which may then be ingested and become even more dangerous.

3. Water exposed to a radioactive element will also become toxic and be potentially harmful to any organisms that come into contact with it.

C. Persistent Organic Pollutants (POPs)

1. Persistent organic pollutants such as DDT and PCBs are synthetic, carbon-based compounds that do not break down easily.

REFERENCES

Dose Response Curve *https://www.sciencedirect.com/topics/ agricultural-and-biological-sciences/dose-response*

Atmospheric Pollution

I. Introduction to Air Pollution

A. Air pollution is caused by a substance added into the atmosphere that can damage the health of ecosystems and the organisms that make them up, especially humans. Air pollution can affect both indoor and outdoor air quality.

B. Point Sources. Many air pollutants are added to the atmosphere by easily identifiable sources. As discussed in Chapter 3, a *point source* is a single, stationary, localized source of emissions. Examples include the smokestacks of factories or power plants, methane emissions from a landfill, or smoke and ash from a forest fire.

C. Nonpoint Sources. Also known as dispersed or mobile sources of pollution, *nonpoint sources* are usually difficult to identify directly as the source for a particular pollution problem. For example, acid rain is a result of nonpoint air pollution. Even though some of the chemicals that combine to make acid rain may have originated from identifiable smokestacks, most pollutants, once released into the atmosphere, travel with wind currents where they mix and chemically recombine. This means that the original source of pollution over a particular area may be hundreds of miles away and impossible to identify directly.

The major theme about air pollution is that combustion (burning) of organic material in any of its forms releases a host of chemicals into the atmosphere, many of which are toxic to living organisms. This includes fossil fuels, wood, and other biomass—even cigarettes! Most of these pollutants aggravate respiratory tissues, resulting in increased rates of asthma, lung cancer, and other respiratory diseases.

D. Major Outdoor Primary Air Pollutants

Primary pollutants are emitted directly from a source, often from combustion processes.

Pollutant	Sources	Human/Environmental Effects	Reduction/Remediation
NO_x: nitrogen oxides, including NO, NO_2, and N_2O	A common product of combustion found in auto emissions, power plant smokestacks, etc. and agriculture (most N_2O comes from agriculture).	Causes respiratory irritation and asthma; combines in the atmosphere to produce tropospheric ozone (a major component of smog) and nitric acid, which may fall as acid deposition	Reduce dependence on combustion to produce usable energy; strict emissions standards for vehicle exhaust; holding coal-fired power plants to higher emissions standards, use of catalytic reduction units.
SO_2: sulfur dioxide	Mostly from burning fossil fuels in coal-burning power plants and factories; volcanic eruptions	Respiratory irritant and aggravates asthma, can lead to bronchitis and emphysema; can combine in the atmosphere to form sulfuric acid leading to acid deposition	Reduce dependence on combustion to produce usable energy; burn higher grades of coal (such as anthracite), which contains less sulfur; control SO_2 in smokestacks by using wet scrubbers that utilize limestone or sea water; use fluidized combustion technology
Particulate matter (PM) of less than 10 μm in diameter	Burning any type of organic materials from fossil fuels to forest fires; dust and dirt from unpaved roads and construction sites	Particles can accumulate in and damage lung tissue; the smallest of these particles can enter a person's bloodstream; aggravates respiratory tissues, which can lead to asthma, bronchitis, emphysema; increases the likelihood of developing heart disease, lung cancer; forms industrial smog	Reduce dependence on combustion to produce usable energy; use electrostatic precipitators in power plants to capture particles before they leave the smokestack
CH_4: methane	A natural product of anaerobic decomposition (including gas from living organisms like cattle and termites); landfills; natural gas leaks	Twenty times more potent greenhouse gas than carbon dioxide; reacts with NO_x to make ozone in the troposphere	The most common way to reduce methane pollution is to trap methane and use it as a source of energy production (natural gas)

II. Secondary Air Pollutants

A. Secondary air pollutants form as a result of chemical reactions among primary pollutants in the atmosphere. As a result, the only way to prevent secondary pollutants is to eliminate the primary pollutants that cause them to form.

B. Acid Deposition. More commonly known as acid rain, acid deposition includes both wet deposition (rain) and dry deposition (aerosols and particles) of acids from the atmosphere.

1. Causes of Acid Deposition. Nitrogen oxides (NO_x) and sulfur dioxide (SO_2) are released into the atmosphere primarily by burning fossil fuels. They react with water and other chemicals in the presence of sunlight to produce nitric and sulfuric acids.

2. Environmental Effects of Acid Deposition

   i. Acidic rainwater. The pH of rainwater is naturally acidic, about 5.6 pH. This is because rainwater naturally reacts with carbon dioxide in the atmosphere to form carbonic acid as it falls. In areas with high levels of industrial pollution, rainwater pH is lowered even further into the acid range as it reacts with sulfur dioxide (SO_2) to form sulfuric acid and nitrogen oxides (NO_x) to form nitric acid.

   ii. Disrupted ecosystems in lakes, streams, and rivers. Organisms in aquatic ecosystems each have a particular range of tolerance for pH. While some species may thrive in acidic environments, most species prefer water that is near to neutral pH.

   iii. Acidification of soil. Acidification leads to the leaching of essential nutrients from the soil. It may also lead to chemical reactions releasing excess aluminum into the soil. Aluminum is a metal that in elevated concentrations is toxic to most forms of life.

   iv. Corrosion of outdoor structures. Acids are corrosive. Even small amounts of acid deposition may, over time, lead to corrosion of concrete, marble, and the metals used in buildings, bridges, and public art.

C. Tropospheric Ozone (O_3)

1. Although ozone forms naturally in the atmosphere, human activities can cause it to become more concentrated in the lower atmosphere. Ozone forms as a product of the reaction of nitrogen oxides with volatile organic compounds (VOCs), such as hydrocarbons in the atmosphere.

2. Ozone is a very corrosive chemical and can destroy plastics, fabrics, nylon, and living tissues. The irritation of plant tissues reduces their ability to perform photosynthesis.

3. Ozone is an ingredient in photochemical smog.

D. Photochemical Smog

1. *Photochemical smog* forms when nitrogen oxides and volatile organic compounds (VOCs) react with heat and sunlight to produce a brown haze.

2. Ozone is a harmful component of photochemical smog. Nitrogen oxides form early in the day, and the formation of ozone reaches peak levels in the afternoon from the reactions with sunlight. Ozone concentrations are highest in sunny, urban areas with a high concentration of vehicles.

3. Photochemical smog can cause respiratory problems and eye irritation. Levels of photochemical smog can be lowered by reducing emissions of nitrogen oxides and VOCs.

4. *Thermal inversions.* Photochemical smog may be intensified by a thermal inversion, where a column of descending cool air traps the smog close to the ground.

   i. Mexico City, Los Angeles, and Beijing have serious problems with this type of air pollution.

III. Indoor Pollutants and Sick Buildings

A. Sources of Indoor Air Pollutants

1. Indoor air pollutants can come from natural sources, human-made sources, and combustion.

Indoor Air Pollutant	Sources	Human/ Environmental Effects	Reduction/Remediation
Mold, fungus, bacteria	These biological contaminants may come from the outside and grow in ventilation ducts, insulation, carpets, ceiling tiles, and so on, especially where heat and moisture are present.	This is a major cause of sick building syndrome; allergic reactions include headaches, coughing, sneezing, dizziness, and respiratory irritation.	Proper maintenance of air filters; increased ventilation; in many cases, the only remedy is to remove and replace all materials that may be contaminated.
Asbestos	In buildings built before 1975, asbestos was commonly used in floor and ceiling tiles, insulation, and other applications.	Tiny, microscopic fibers that are breathed can cause lung cancer 20 or more years after exposure.	In 1975, the use of asbestos was banned; in older buildings, asbestos may need to be removed by professionals or left in place and covered over.

Indoor Air Pollutant	Sources	Human/ Environmental Effects	Reduction/Remediation
Radon	A natural decay product of U-238 in bedrock, it is found in almost all water, soil, and rock. It enters homes through cracks and opening in floors and walls.	Lung cancer (second only to smoking in causing lung cancer deaths).	Ventilation is the most reliable way to reduce radon in the home; special home ventilation systems draw radon up from the soil and send it out above the house.
Carbon monoxide (CO)	Improperly ventilated fireplaces; leaks from gas appliances like water heaters, stoves, and clothes dryers; the tailpipe of a car in a closed garage.	CO is an odorless, invisible gas that can kill you before you are even aware it is there. CO binds to hemoglobin in the blood, and keeps it from picking up and carrying oxygen to the body's cells; it essentially leads to suffocation.	Every home that has gas appliances and/or wood-burning fireplaces needs to have carbon monoxide detectors; the detectors need to be checked regularly. Gas appliances need to be professionally installed and vented.
Volatile organic compounds (VOCs)	VOCs are found in many products: new carpet, paint, foam padding in furniture, and household cleaners, to name a few.	Cause a wide array of harmful health effects, including eye and respiratory irritation, headaches, and dizziness; increased risk of cancer.	Many products are now made without VOCs. Properly ventilate your house—open windows and doors to increase airflow. Properly store and dispose of chemicals, including paints and solvents.

IV. Reduction of Air Pollution

A. Strategies to reduce air pollution include regulatory practices, conservation, and alternative fuels.

1. Vapor recovery nozzles are installed on gas pumps. They capture gasoline vapors while pumping gas and prevent them from entering the atmosphere. A vapor recovery nozzle is an air pollution control device on a gasoline pump that prevents fumes from escaping into the atmosphere when fueling a motor vehicle.

2. Catalytic converters are installed in the exhaust system of internal combustion engines. They convert harmful pollutants such as hydrocarbons and NO_x into less harmful molecules such as water, nitrogen, and oxygen.

3. Dry and wet scrubbers are air pollution control devices installed in industrial exhaust systems to remove particulates and acid gases such as SO_x. Scrubbers do not remove NO_x.

 ## V. The Clean Air Act

A. By 1955, federal legislation was needed in the United States to reduce air pollution. Regulations were issued, which, in 1963, became the Clean Air Act. Its provisions include:

1. Authorization and funding studies of air quality to learn more about the presence and effects of atmospheric pollutants.

2. Setting enforceable regulations to limit emissions from stationary sources such as factories and power plants, as well as mobile sources, e.g., cars, trucks, and ships.

3. Developing programs to monitor and reduce acid deposition and the primary pollutants that cause its formation.

4. Establishing a program to phase out the use of chemicals that deplete stratospheric ozone.

5. Establishing a cap and trade program for SO_2.

   i. Cap and trade is a system in which maximum allowable emissions are set for each industry.

   ii. Businesses that can reduce their emissions below the standards are awarded credits that they can sell to other businesses that cannot meet the limits.

   iii. This makes economic sense because it creates a huge financial incentive for industries to reduce emissions quickly. Allowing the sale of emissions credits means that many groups can rapidly recover the costs associated with reducing emissions.

VI. Stratospheric Ozone Depletion

A. *Ozone* in the upper atmosphere, or stratosphere, is essential to life. In the 1970s, scientists first detected a reduction in stratospheric ozone. This resulted in legislation to phase out ozone-depleting chemicals worldwide.

B. "Good" Ozone and "Bad" Ozone. As you read earlier in this chapter, ozone in the troposphere is a powerful pollutant. In the stratosphere, however, it is an essential blocker of excess ultraviolet (UV) radiation.

C. Chemicals that Deplete Ozone. Compounds that break down to release chlorine or bromine are the major contributors to ozone depletion. Many of these compounds are used as refrigerants, coolants, aerosol propellants, and industrial solvents. Fluorine does not contribute to ozone depletion.

   1. Chlorofluorocarbons (CFCs or CCl_2F_2) are also greenhouse gases. In the 1970s, the research of chemists Roland and Molina showed the negative effects of CFCs on stratospheric ozone.

   2. Halocarbons and halons are a type of CFC used as fire retardants, pesticides, and foam insulation.

D. Health and Environmental Effects. Excess UV radiation in the lower atmosphere causes great damage to the cells of living organisms. Aside from the increasing incidence of sunburn, UV radiation also greatly increases the human risk of skin cancer and cataracts. Plant photosynthetic productivity is also greatly reduced due to cellular damage. Animals such as fish, amphibians, and other organisms lacking protective fur or scaly skin are also harmed directly.

E. The Montreal Protocol proved that global problems can be solved by a combination of international cooperation and scientific advances. It requires participating nations to phase out the use of ozone-depleting chemicals in favor of less harmful alternatives.

Test Tip

Beware! *Many students confuse the issues of ozone depletion and global climate change. This is further complicated by the fact that many ozone-depleting chemicals are also greenhouse gases. Make sure you separate the two issues in your mind. Reduction in stratospheric ozone does not increase global temperatures. Ozone depletion leads to increased UV penetration, resulting in increased rates of cancer for humans and loss of photosynthetic productivity in plants.*

VII. Noise Pollution

A. Noise pollution is exactly what it sounds like. High levels of noise can cause physiological stress or hearing loss in humans.

   1. Noise pollution can make it difficult for animals, especially birds, to communicate or hunt, causing changes in migratory routes.

   2. Sources of noise pollution in urban areas include transportation and construction sites.

Aquatic and Terrestrial Pollution: Aquatic Pollution

Chapter

15

I. Major Sources of Water Pollution

A. **Point Sources.** Like atmospheric pollution, point sources are easily identifiable, localized sources of pollutants—the drainpipe from a factory, a landfill, a sewage treatment plant, or a fish farm are all examples of point sources of water pollution.

B. **Nonpoint Sources.** Nonpoint sources are more difficult to identify and, in the case of water pollution, currently the greatest problem.

1. The major source of nonpoint pollution is *runoff*. Agricultural runoff may contain animal wastes, excess fertilizer, and toxic chemicals used as herbicides and pesticides.

2. On a smaller scale, runoff from neighborhood landscaping may also contain many of these pollutants. Roads and parking lots often lead to runoff contaminated by motor oil, antifreeze, and other chemicals used on and in motor vehicles.

II. Water Quality Tests

Parameter	Description/ How It Is Tested	Desired Results	Environmental Significance
Dissolved oxygen (DO)	The amount of oxygen dissolved in a sample of water. Tested with a variety of methods, including Winkler titration and digital probes. *Note:* The temperature of the water is important. Cold water has a higher capacity to hold DO than does warmer water.	High DO readings, usually in the range of 7–9 ppm or above, are best. Few fish can survive in water of less than 3 DO ppm. Low DO levels put aquatic organisms at risk of hypoxia (lack of oxygen).	All aerobic organisms require oxygen for cellular respiration. Aquatic organisms rely on oxygen dissolved in the water to satisfy this need. DO is higher in flowing water due to interaction between the surface water and oxygen dissolved in atmospheric air.

(continued)

Parameter	Description/ How It Is Tested	Desired Results	Environmental Significance
Biological oxygen demand (BOD)	Samples of water are taken in clear bottles. One is darkened to prevent sunlight from entering; one is left open to light penetration. After a set period, the difference in dissolved oxygen in the two bottles is used to determine the BOD.	Lower BOD ratings are best. Most healthy aquatic ecosystems have a BOD of less than 10 ppm.	Closely related to DO testing, BOD indicates the levels of aerobic bacterial activity in a water sample. High BOD signals that there is a source of organic material to feed decomposing bacteria and encourage their population growth.
pH	A measure of the acidity or alkalinity of water. Using pH indicator paper, pH solution, or a digital probe.	Most aquatic organisms prefer a pH near neutral (7). The pH scale is logarithmic, meaning that a pH of 6 is 10 times more acidic than a pH of 7. (A pH of 5 is 100 times more acidic than 7!)	Low pH may indicate acid deposition and have harmful effects on aquatic life.
Hardness/ alkalinity	Alkalinity is a measurement of the buffering capacity of water; it is the ability of a body of water to neutralize acids without changing the pH of the water. Hardness is a measurement of the concentration of metal ions like Ca^+ and Mg^+.	Higher alkalinity is good in aquatic ecosystems; it acts as a buffer against pH drops from acid deposition. Zero to 60 ppm is soft water; above 60 ppm is in the hard-water range.	Alkalinity is associated with hard water because the calcium and magnesium ions that make water "hard" are great acid buffers. Hard water is mostly a problem in domestic use. It can leave behind cloudy deposits on glassware and requires more detergents to wash clothes or dishes.
Turbidity/ total suspended solids (TSS)	In the field, a Secchi disc is lowered into the water until it can no longer be clearly seen. In the lab, a spectrometer is used.	Low turbidity is best.	Turbid water has low visibility. As turbidity increases, the euphotic zone decreases because sunlight is blocked from penetrating deeper water by the suspended particles in the water.

Parameter	Description/ How It Is Tested	Desired Results	Environmental Significance
Fecal coliform	A sample of the water is placed in a culture dish and incubated to see if any fecal coliform bacteria colonies grow.	Zero colonies Presence of fecal coliform in a water sample is an indication that the water is likely to be contaminated by mammal or bird feces.	Although these bacteria, alone, are not harmful, they are indicators of the presence of feces, which may carry a host of pathogens that *are* harmful to human health, such as viruses, parasites, and harmful bacteria.
Nitrates and Nitrites	Nitrogen exists in the environment in many forms and changes forms as it moves through the nitrogen cycle. Nitrate levels can be assessed using a meter, probe, or chemical analysis kit.	Under 10 mg/L is required for drinking water.	Excessive concentrations of nitrate-nitrogen or nitrite-nitrogen in drinking water can be hazardous to health, especially for infants and pregnant women. Runoff of these nutrients can lead to algal blooms and decreased DO levels.
Phosphates	Phosphorus is an essential plant nutrient. The amount of phosphorus is often reported as the amount of phosphate ion (PO_4^{3-}) and is measured with a meter, probe, or chemical analysis kit.	Phosphate levels as low as 0.15 mg/L are considered sufficient to trigger algal blooms in surface waters, and levels as low as 0.5 mg/L are considered unsafe for drinking.	High concentration of phosphates in surface water are linked to algal blooms and decreased concentration of DO.

Test Tip

Remember that one of the FRQs on the exam is about experimental design. You might be asked to design an experiment to determine if an aquatic ecosystem has been impacted by human activity. If your experimental question is to determine if water had been impacted by overflow from a sewage treatment plant, you would test for fecal coliform.

III. Process of Eutrophication

A. Eutrophication

1. Eutrophication happens when a body of water has become enriched with nutrients and minerals. But too many nutrients can result in a harmful depletion of oxygen.

2. When eutrophication happens naturally over a long period by gradual accumulations of sediment and other organic materials in an aquatic ecosystem, it leads to well-developed, biodiverse ecosystems.

B. Oligotrophic Lakes

1. When a new lake is formed, it is often just a depression in the ground that fills with water through rainfall or flooding of an existing body of water.

2. In most cases, the new lake is little more than bare rock covered with water, and with few nutrients, it has few organisms.

3. These new, low-nutrient lakes are called *oligotrophic*.

C. Natural Eutrophication

1. In a natural system, nutrient-rich sediments containing nitrogen and phosphorus build up over time on the bottom and around the edges of an oligotrophic lake, gradually making more plant life a possibility. These sediments ultimately increase the natural biodiversity of the ecosystem as more nutrients make more life possible.

2. In most ecosystems, nitrogen and phosphorus are the primary limiting factors because these two chemicals are essential for building the molecules of life, such as DNA, RNA, amino acids, and phospholipids.

3. When present in low concentrations, nitrogen and phosphorus stimulate the growth of plants and, as a result, other aquatic organisms that rely on plants for food.

D. Cultural Eutrophication

1. When the accumulation of nutrients happens rapidly as a result of anthropogenic (human-caused) activities, the ecosystem is usually harmed. This is called *cultural eutrophication*.

2. Runoff. Agricultural or other nutrient-rich runoff enters a body of water. An excess of nitrogen and phosphorus can lead to algal blooms. Both nutrients are commonly found in runoff from fertilizers used in agriculture and residential landscaping, as well as sewage effluent. Because these nutrients are limiting factors, adding them into the system in excess stimulates rapid growth of plant life, especially algae.

3. Algal blooms. When concentrations of nutrients in an aquatic ecosystem increase, it causes excessive growth of algae, an algal bloom. Algae are the main producers in the ecosystem, and they grow and reproduce rapidly with a plentiful supply of nutrients. When the algae die, microbes (decomposers) digest the algae and deplete the oxygen. When a waterway has a low level of dissolved oxygen, it is referred to as hypoxic. These conditions can result in large die-offs of fish and other aquatic organisms.

E. Reversing Cultural Eutrophication

1. Anthropogenic causes of eutrophication include fertilizer (nitrogen and phosphorus) runoff from lawn and agricultural fields and overflow from wastewater treatment plants.

2. Herbicides/algaecides may be added to the water to kill the algae, but unless measures are taken to remove excess nutrients from the water, this is usually only a temporary solution.

3. Pumping oxygen into the water. As a quick fix to combat low DO while more permanent steps are being taken, oxygen gas can be pumped down into the deeper depths of the water to boost DO levels so fish and other aerobic organisms avoid hypoxia. Again, this is not a permanent solution.

4. Dredging lakes to remove accumulated detritus. The buildup of decaying plant material is a major contributor to the drop in DO as decomposing bacteria populations grow in response to this drastic increase in food supply. By the removal of the detritus, bacterial populations can be kept in check. This is one step toward a permanent solution.

5. Bioremediation to remove excess nutrients. The best way to avoid cultural eutrophication is to bring water nutrients to normal levels. First, the source of the runoff or effluent must be located and stopped. Next, certain species of fast-growing plants (duckweed) can be used in conjunction with some or all of the methods men-

tioned above to pull excess nutrients out of the water. Duckweed is convenient because it is a commonly found aquatic plant (not likely to be an exotic) that floats on top of the water and can be scooped out once it has done its work. Any remaining duckweed is kept in check by lower nutrient levels in the remediated water.

Make sure you know the entire process of eutrophication. This extremely important concept shows up on every exam, usually relating to more than one question. Understanding how all the factors (DO, algal blooms, and fish kills) are connected is the key.

IV. Municipal Wastewater

A. Definition

1. Municipal wastewater is water discharged by domestic residences, businesses, industry, and stormwater.

2. Wastewater is treated in wastewater treatment plants, then either re-used or returned to nature.

B. Steps of Wastewater Treatment

1. Primary treatment = physical removal.

   i. Water is passed through a series of screens followed by settling of waste to the bottom of a tank.

2. Secondary treatment = biological process.

   i. Water is then moved to large aerated ponds, and decomposing aerobic bacteria break down the organic matter into carbon dioxide and inorganic sludge. This sludge is now a waste product that must be disposed. It can be spread on agricultural fields, sent to a landfill, or converted into biosolid (fertilizer) pellets.

   ii. After exposure to the bacteria, water is treated with disinfecting chemicals, including chlorine or ozone, or UV lights to kill any living organisms present.

3. Tertiary treatment = further purification.

   i. Nutrient removal. Using advanced filters or even treatment wetlands, excess nutrients like nitrates and phosphates are removed from treated wastewater.

4. Treated wastewater can be returned to the environment by being released into a local waterway. In some large cities with water shortages, it may be further treated for purification and returned into the municipal water supply to be reused in households, industry, and public works.

C. Is Treated Wastewater "Clean"?

1. Treated wastewater is considered "clean" when it tests within acceptable levels for nitrates, phosphates, bacteria, and disinfecting chemicals such as chlorine.

2. Other potentially harmful chemicals, like pharmaceuticals, including antibiotics and human hormones used in birth control pills, are not removed or tested for. Many of these are endocrine disruptors that can result in cancerous tumors and birth defects.

V. Other Water Quality Concerns

A. Heavy Metal Contamination

1. Heavy metals are extremely toxic to humans in relatively small concentrations.

   i. In aquatic ecosystems, heavy metals tend to accumulate in large predatory fish due to biomagnification. As a result, people should limit their intake of these fish in their diet.

   ii. The neurotoxic effects of heavy metals are not limited to humans. Aquatic organisms also suffer a wide array of health problems in water with elevated levels of metals such as lead, mercury, zinc, and arsenic.

2. Mercury, and to a lesser extent arsenic, are products of the combustion of coal to produce energy. They end up in aquatic ecosystems as small airborne particles that are carried in the wind and eventually settle to the ground or in bodies of water either as dry deposition or mixed with rainwater. These particles are converted into methyl mercury.

3. In addition to being directly harmful to organisms, acidification of water actually makes heavy metals in sediments more soluble, thereby increasing concentrations of zinc, mercury, and other heavy metals in the acidified water. This drop in pH may be caused by acid deposition or also by acid drainage from coal-mining operations.

B. Oil Spills

1. Oil spills can kill marine organisms that have been exposed to hydrocarbons. The oil can coat the feathers of birds and fur of marine mammals, which can lead to hypothermia. In addition, the animals will often ingest the oil in an attempt to remove it.

   i. Common methods of cleaning up oil spills include mechanical containment, chemical methods, and biological agents.

      ➤ Mechanical containment. Large booms filled with absorbent materials (even human hair!), skimmers, and other types of barriers are placed in the water to attempt to contain the spread of oil.

      ➤ Chemical methods. Detergents and other dispersing agents attempt to break down oils, while clumping or gelling agents cause the oil to form large globules that can be more easily scooped or skimmed from the water.

      ➤ Biological agents. Naturally occurring oil-consuming bacteria will, over time, eventually break down the oil. To dramatically hasten the process, water may be fertilized with nutrients such as nitrogen and phosphorus to encourage bacteria growth. Water may also be seeded with oil-consuming bacteria to boost populations.

VI. Water Quality Legislation

A. Clean Water Act (CWA)

1. Originally passed in 1972 and revised several times since, the mission of the CWA is to restore and maintain the chemical, physical, and biological integrity of America's waterways to support wildlife and recreational activities. The provisions of the CWA are intended to reduce and prevent both point and nonpoint sources of water pollution in the nation's surface waters. There are no provisions specific to groundwater protection.

B. Safe Drinking Water Act (SDWA) of 1974

1. The focus of the SDWA is to maintain the purity of any water source that may potentially be used as a source of drinking water. This includes both surface water and groundwater.

Federal environmental law doesn't appear very often on the AP® exam, but when it does, it's usually in a free-response question. If asked to discuss a law about safe water, you can rely on either the Clean Water Act or the Safe Water Drinking Act. If wildlife or aquatic ecosystems are involved, choose the CWA. If human health is a focus of the question, then the SWDA is a better bet. Both laws are very broad and cover a variety of circumstances.

REFERENCES

USGS Water Quality. *http://water.usgs.gov*

Biological Agents/Emergency Management/U.S. EPA. *http://www.epa.gov/emergencies*

Aquatic and Terrestrial Pollution: Solid Waste Disposal

Chapter

16

I. Municipal Solid Waste

A. Solid waste is discarded material that is not liquid or gas and comes from industrial, agricultural, business, and domestic sources.

B. Paper is the leading type of municipal solid waste produced (by weight) in the U.S. More paper is thrown away than any other material.

C. Options for the Disposal of Solid Waste

1. Landfills

   i. Landfills are the most common disposal method in the U.S. for solid waste. Such waste can contaminate surrounding soil and groundwater. Moving through layers of contaminants, the waste can mix with the groundwater. This is known as leachate. Also, the decomposition of organic matter in the landfill can generate harmful methane gas.

   ii. Sanitary landfills are designed to isolate waste from the environment. These landfills contain a bottom layer composed of plastic or compacted clay to protect groundwater. Sanitary landfills also have leachate and methane collection systems to protect humans and the environment from harmful components of the waste. The methane can be burned to heat water, create steam, turn a turbine, and generate electricity.

   iii. Waste is often crushed or compacted before being placed in landfills. Crushing trash often creates smaller particles of waste with a higher surface area. This is a problem because it increases the potential for leaching.

   iv. Some items, such as rubber tires, cannot be disposed of in sanitary landfills and are often dumped illegally. Improperly disposed of tires collect water and serve as breeding grounds for mosquitoes that transmit infectious diseases.

v.   Dumping waste in oceans damages ecosystems and humans. Marine animals can become ensnared in the plastic, and they can also ingest it. Humans can then be exposed to the plastics when consuming the fish.

vi.   When landfills reach capacity, they can be capped and turned into parks and recreational areas.

2.   Incineration

i.   *Incineration* is the burning of waste.

ii.   A waste-to-energy facility removes all materials that can be recycled and recovers energy from the combustion of the remaining waste. Much of a waste-to-energy facility is dedicated to air pollution control devices that remove ash and acid gases from the emissions.

3.   Composting

i.   Any biodegradable material is a candidate for composting. A material is considered biodegradable if exposure to natural elements, such as water, sunlight, and bacterial decomposition, can break it down. Composting can be done on a large or small scale.

ii.   Individual homeowners often choose to do backyard composting by collecting kitchen scraps and yard waste in a pile and using the finished compost as a rich organic fertilizer.

iii.   Composting can also be industrially scaled. More municipalities are using large, pressurized digesters to rapidly break down organic wastes. The resulting compost may then be used as a layering agent in landfills or spread as fertilizer.

4.   Recycling

i.   *Recycling* is the collection and recovery of useful materials that can be remade into new products.

ii.   Pros of Recycling

➤   Recycling keeps many types of waste out of landfills.

➤   Materials like glass and most types of metal can be recycled many times without much loss in the purity or usefulness of the materials.

➤   Recycling reduces the need for environmentally damaging mining and extraction of metals like bauxite (for aluminum) and copper.

➤ Recycling paper reduces deforestation for wood pulp.

➤ When compared to the energy required to extract, purify, and process many raw materials, especially metal ores, recycling is much more efficient and, in many cases, produces fewer pollutants.

iii. Cons of Recycling

➤ Recycling is energy- and often water-intensive. In areas where water is limited, processes like recycling paper may be difficult to justify.

➤ In most cases, recycling a material means that it must be melted and re-formed into a new product. In addition to requiring large inputs of energy, the melting process may release toxic gases into the atmosphere or result in harmful chemical by-products that must be disposed of.

➤ Recycling can be expensive. Consider the costs of providing recycling bins to every household, maintaining and powering the collection vehicles, paying workers to collect and process materials, and maintaining the recycling facility. If there is not a high demand for the resulting recycled products, municipalities often must use tax dollars to support the process.

➤ *Downcycling* is when the products made from recycled materials are of a lesser quality or purity than the original. Some materials, most notably plastics and paper, are more prone to downcycling.

— For example, used paper often contains inks and dyes used in the printing process. It may be impossible to remove all of the impurities when making new paper, so writing paper and newsprint is often downcycled into products like cardboard or paperboard. Because plastics come in so many varieties, they are often mixed together when melted, so the resulting re-formed plastic is a combination that is of a lower grade than the original(s).

5. Integrated Waste Management

i. Most municipalities use a combination of disposal methods to manage their waste.

ii. Where appropriate, recycling is encouraged or enforced. Some materials are incinerated or composted, and the remainder is placed in a sanitary landfill.

6. Reducing and Reusing

   i. The familiar three R's are reduce, reuse, and recycle. Reducing and reusing help to keep waste from being generated in the first place. In terms of environmental sustainability, reducing and reusing are the best options.

   ii. *Reducing* simply means to throw away less.

   ➤ Buy products with less packaging. Bring reusable grocery bags to the store instead of using the disposable plastic ones. Avoid buying and using paper plates, napkins, and any disposable products. Use a metal water bottle that you fill from the tap instead of buying bottled water.

   ➤ Reducing helps in several ways: Landfill space is saved by the reduction in trash, fewer materials need to be mined or extracted from the environment, and the consumer saves money that would otherwise be spent on disposable products.

   iii. *Reuse* means to think of another way to use the item.

   ➤ Instead of throwing something away, think of other ways it could be put to use. Donate unwanted items to thrift shops or shelters. Use "trash" for art projects: egg cartons make great paint trays; broken dishes are good for mosaics; cardboard boxes can be made into a child's playhouse, and so on.

II. Hazardous Waste

A. *Hazardous waste* is materials that are known to be flammable, corrosive, toxic, or reactive.

1. Electronic waste (e-waste) includes discarded computer monitors, televisions, and cell phones. This waste may contain heavy metals such as mercury. These metals can leach from landfills into groundwater where they can impact humans and ecosystems. Many communities have implemented e-waste recovery programs to ensure proper disposal.

B. Waste Disposal Legislation (CERCLA and RCRA)

1. The Comprehensive Environmental Response, Compensation, and Reliability Act (CERCLA)

   i.   Also known as the Superfund Act, CERCLA was enacted to handle industrial contamination in sites where no direct individual or party could be held responsible for cleanup.

   ➤ For example, a company released arsenic into the soil many years ago and then went out of business. The site was sold to a new owner. Who is now responsible for cleaning up the arsenic? (Probably Uncle Sam!)

2. The Resource Conservation and Recovery Act (RCRA)

   i.   Commonly called the "cradle to grave" act, RCRA sets specific regulations in the manufacture, transport, storage, use, and ultimate disposal for a host of hazardous chemicals. Its major provision requires extensive documentation at every step to ensure that hazardous wastes are disposed of properly.

Global Change: Global Climate Change

Chapter

17

I. The Greenhouse Effect

A. The *greenhouse effect* occurs when gases in the lower atmosphere (troposphere) trap infrared radiation from the sun. This heat radiates back to Earth and further warms it.

B. The major greenhouse gases are listed in the following table.

Greenhouse Gas	GWP*	Major Sources	Remediation or Reduction
Water vapor (H_2O)	2	Nonanthropogenic (natural). Warming naturally increases atmospheric water vapor due to evaporation.	Although water vapor is a greenhouse gas, it is not a major factor in climate change because it has a short residence time in the atmosphere.
Carbon dioxide (CO_2)	1	Burning fossil fuels.	Move energy generation from fossil fuels to other sources such as hydro, wind, solar, nuclear. Carbon sequestration.
Methane (CH_4)	20	Livestock. Anaerobic respiration from decomposition in swamps and landfills.	Capture, compress, and burn as natural gas. CH_4 degrades quickly in the air so reducing emissions would have an almost immediate effect.
Nitrous oxide (N_2O)	300	Agriculture	Use less nitrogen fertilizers, minimize soil tillage

(continued)

Greenhouse Gas	GWP*	Major Sources	Remediation or Reduction
Chlorofluoro-carbons (CFCs) (CCL_2F_2)	6500	Not usually found in nature; purely anthropogenic. Refrigerants.	Montreal Protocol banned CFCs. CFCs persist in the atmosphere for hundreds of years, so even though they are no longer being released, they will continue to affect the atmosphere.
Ozone (tropo-spheric) (O_3)	Potent, but no specific number	A component of photochemical smog as a result of automobile emissions, burning biomass, and industrial pollution.	Reduce emissions of smog-causing gases (NO_x, SO_2) and particulate matter by using catalytic converters on cars and smokestack scrubbers.

*GWP = Global warming potential in a scale that allows for an evaluation of the impact of different greenhouse gases. Carbon dioxide is given a value of 1 as a reference for comparison to other greenhouse gases. The greater the number, the higher the potential to trap heat in the atmosphere.

Source for GWP: National Climate Data Center (*http://www.ncdc.noaa.gov/oa/climate/globalwarming.htm*).

Expect the AP® exam to address, in some way, the fact that the burning of fossil fuels and biomass are the major sources of increased carbon dioxide in the atmosphere.

II. Focus on Carbon

A. Fossil Fuels

1. All life on Earth contains carbon.

2. Fossil fuels originate from carbon compounds naturally released from living things as they slowly decay over long periods of time and under intense pressure.

   i. Petroleum (oil) comes from the remains of prehistoric plants and animals. Coal comes from the plants in and around ancient swamps.

B. Causes of Seasonal Fluctuations of Carbon

1. Every year there is a measurable cycle of carbon dioxide caused by seasonal changes in photosynthesis in the world's forests. This cycle is represented by the dashed line on the graph. The dotted line is the mean average for each year.

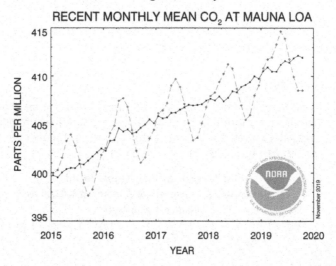

RECENT MONTHLY MEAN CO_2 AT MAUNA LOA

2. Photosynthesis allows plants to take in carbon dioxide and, in the presence of sunlight, manufacture glucose and oxygen. Therefore, more photosynthesis lowers atmospheric CO_2.

3. In the cold winter months (or dry season near the equator), trees in deciduous forests lose their leaves, and evergreens go into a dormant state to conserve water. This leads to a significant, world-wide drop in photosynthesis during winter in the northern hemisphere because there is much more land and vegetation north of the equator. As a result, atmospheric CO_2 goes up.

4. For the same reason, there is a measurable, worldwide decrease in atmospheric CO_2 during the warmer months (or wet season) in the northern hemisphere as photosynthesis reaches its peak.

C. Burning Fossil Fuels

1. Burning fossil fuels increases atmospheric carbon by releasing carbon that had been sequestered underground in coal and petroleum.

D. Clear-Cutting

1. Clear-cutting a tropical rainforest increases anthropogenic climate change because of a reduction in CO_2 uptake by plants (from photosynthesis); also, burning the logged materials releases carbon into the atmosphere.

III. Evidence of Global Climate Change

A. Ice Core Data

1. Data from Antarctic ice cores measure atmospheric gas composition from hundreds of thousands of years ago. Ice core data shows a distinct correlation between atmospheric CO_2 concentration and air temperature. See the following graph.

The Vostok (Antarctica) Ice Core Record
Carbon Dioxide versus Temperature for the last 420,000 years

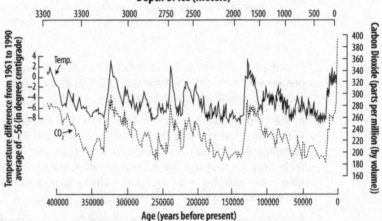

2. For the majority of the graph, CO_2 and temperature follow a very similar pattern.

3. Only at the very far right (near the present time) do we see a discrepancy: CO_2 concentration is sharply increasing to ever-higher levels than anywhere else on the graph (nearly twice as high).

4. During this same time, temperatures are at the high end of the cycle. The current concern is that, because CO_2 is a greenhouse gas, the rise in CO_2 is likely to lead to a further rise in temperature

that will be higher than any temperature recorded on Earth in the last 400,000 years.

5. However, correlation does not equal causation. Scientists cannot be absolutely sure if increased CO_2 is causing the rise in temperature. It is possible that the rise in temperature may be causing the rise in CO_2, or perhaps there is some other factor that hasn't yet been discovered. However, because CO_2 is a known greenhouse gas, there is a very high likelihood that a significant rise in temperature will result from a spike in atmospheric CO_2 concentration.

B. Historical Data

1. Data shows that global temperatures have increased by 0.5°C in the last 100 years.

2. The data also show that the rate of increase for temperature continues to increase.

3. Within the next 100 years, global temperatures could increase by as much as 4°C.

C. Effects of Climate Change on Ecosystems and the Populace

1. Changes in ocean levels. As the atmosphere and oceans warm, a significant rise in the overall sea level will be seen.

   i. The major cause of sea-level rise is thermal expansion of oceans. As liquids warm, their molecules are energized and thus experience more energetic collisions, causing them to spread out. This leads to a slight overall increase in volume that may not be noticeable in a small pan of water, but in an area as large as the Earth's oceans, this will be one of the major causes of sea-level rise.

   ii. Melting ice is another significant cause of sea-level rise.

      ➤ This is true in the case of land-based ice such as mountain glaciers, the Greenland Ice Shelf, and Antarctic continental ice. As this ice melts, the water runs from land to sea, resulting in more ocean water and a sea-level rise.

      ➤ This is *not* true in the case of sea-based ice. Sea-based ice such as icebergs and most of the Arctic ice, do not affect sea level because they are already in the water. As they melt, they just become part of the water in which they were floating (like ice cubes in a glass of water— the glass does not overflow as the ice cubes melt).

iii. Small increases in sea level will cause flooding of coasts and low-lying areas. In estuaries and coastal ecosystems, this will lead to the loss of wetlands, marshes, and intertidal zones that may be important breeding or shelter grounds for birds, fish, and shellfish. This could cause the decline or total loss of these species.

iv. Roughly 40 percent of the world's population lives within 60 miles of a coastline. Rising water will severely affect coastal communities because of the following factors:

> ➤ Loss of property and property values as homes flood and land erodes from increased storm surges and loss of beach sand or mangrove stands.

> ➤ Estuary flooding, which could lead to the loss of fish and shellfish species. Humans who rely on them as a source of income or food will suffer.

> ➤ Loss of beach and fishing areas may lead to the collapse of tourism in coastal areas—a major source of income and jobs.

> ➤ Intrusion of saltwater into water supplies for drinking and irrigation can leave communities with no source of freshwater.

> ➤ Increased conflict because people displaced by flooding will migrate inland into already crowded areas.

2. Large river systems like the Colorado River in the U.S. or the Yangtze River in China could experience the following:

i. Warmer temperatures will increase spring melt of mountain snow pack. For a time this will increase the flow of water in the river (which is good because many people rely on these large rivers for drinking water, hydroelectric power, etc.). In the long term, however, the increased spring melt reduces the overall size of the mountain glaciers, and eventually there will be no snow pack left to supply the river.

ii. Depending on the location of the river, global climate change will likely mean either increased or decreased precipitation.

> ➤ More precipitation would increase the input to surface water and groundwater (good), while runoff from the rain falling on land and running into the river would lead to increased erosion and sedimentation (bad).

➤ Less precipitation would do just the opposite—reduce input to surface and groundwater (bad) and decrease erosion and sedimentation (good).

iii. An increase in the frequency and severity of storms is also projected. Stronger storms would lead to more sedimentation and flooding, and greater runoff.

3. Biome loss. As a result of global climate change, many biomes will no longer exist where they are currently found.

   i. Biomes are shaped by temperature and precipitation. As global patterns of both change, many of the warmer biomes will expand outward toward the poles. This means that the tropics could expand into the southern U.S., while the temperate zones will shift northward into the region that is now boreal forests.

   ii. Earth's atmosphere will contain more water vapor as surface water is warmed and evaporates. This will cause changes in the distribution and frequency of precipitation, and such changes are difficult to predict.

   iii. Specialist species with a narrow range of tolerances would likely become extinct when their fragile ecosystems change too rapidly for them to adapt.

   iv. The distribution of many insect species will grow as warmer climates and increased rainfall means more ideal breeding conditions and less winter die-off. Insect vectors transmit many infectious diseases, such as malaria and West Nile virus. Expanded distributions of mosquitoes and other insect vectors will allow these diseases to spread to currently unaffected areas such as those in the northern latitudes. As insect populations expand, there is also more potential for crop damage from increased pest attacks.

4. Ocean warming. As the atmosphere warms from human activities, so does ocean temperature. This can cause loss of suitable habitat and interfere with the metabolism of many marine organisms. A warmer ocean also causes *coral bleaching* where the algal symbiont is ejected from the coral, leading to loss of color in the coral colony.

5. Ocean acidification. Another consequence of increased greenhouse gases is that more CO_2 is dissolved in oceans. The CO_2 reacts with water to form carbonic acid as illustrated in the equation below.

$$CO_2 + H_2O \rightarrow H_2CO_3$$

Ocean acidification also damages coral reefs because it becomes difficult for the corals to form their skeletal structure.

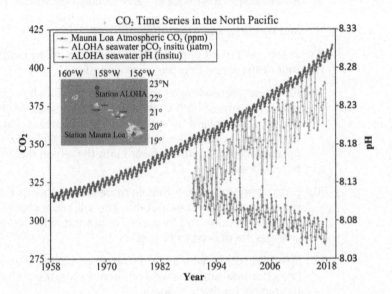

D. A decline in albedo (a measure of reflectivity) is an example of a positive feedback loop that works to accelerate global warming, especially at Earth's poles.

1. Because ice is reflective and light in color, it has high albedo and reflects most of the sun's radiation (heat) away from Earth's surface. Warmer temperatures cause ice to melt, allowing more sunlight to fall on the underlying land or water instead.

2. Land and water are both darker in color and less reflective than ice, so more of the heat is absorbed and less is reflected (low albedo). This leads to an increase in temperature, which in turn accelerates the melting of ice.

3. The albedo effect is a positive feedback loop because the increased temperature leads to even higher temperatures. This is often referred to as a vicious cycle because once begun, it is difficult to halt or reverse.

IV. Reversing the Warming Trend—Laws and Treaties

A. The Kyoto Protocol

1. The Kyoto Protocol was developed at a world summit in 1997 to address reduction of greenhouse gas emissions. But the U.S. rejected it, and soon after the whole movement lost support.

B. Paris Climate Agreement

1. The Paris Climate Agreement was ratified by 185 countries in 2016. Its purpose is to limit global temperature increase and thus the impact of climate change. Each signatory country sets standards for reducing greenhouse gas emissions. The United States withdrew from the Paris Agreement in 2017.

Math Practice

1. *A typical family in Florida uses 2,000 kwh per month of electricity during the hot summer months. All of their electricity is generated by their town's coal-fired power plant. This power plant emits 100 kg of CO_2 per MMBTU (million BTUs) generated at the plant. For every kwh of electricity delivered to a home, 3,412 BTUs of energy are generated at the plant. Use this information to determine how many Kg of CO_2 are emitted per month to generate electricity for this family.*

2. *Refer to the earlier graph titled "Recent Monthly Mean CO_2 at Mauna Loa." Determine the percentage increase in CO_2 from 2006 to 2010.*

ANSWERS:

1.
$$\frac{? \text{ kg CO}_2}{\text{month}} = \frac{100 \text{ kg CO}_2}{\text{MMBTU}} \times \frac{\text{MMBTU}}{1 \times 10^6 \text{ BTU}} \times \frac{3412 \text{ BTU}}{\text{kWh}}$$
$$\times \frac{2000 \text{ kWh}}{\text{month}} = \frac{682.4 \text{ kg CO}_2}{\text{month}}$$

2.
$$\frac{388 - 381}{381} = \frac{7}{381} = 0.018$$
$$0.018 \times 100\% = 1.8\%$$

Based on past AP® exams, the most important things from this section to remember are:

• Burning fossil fuels is the major source of atmospheric carbon emissions.

• The rise in ocean levels from global warming will occur primarily because of thermal expansion of ocean water and not from the melting of land-based ice. (Melting ocean ice will also not significantly change sea levels.)

REFERENCES

Seasonal fluctuations in CO_2 (Mauna Loa) graph: Dr. Pieter Tans, NOAA/ESRL *https://www.esrl.noaa.gov/gmd/ccgg/trends/data.html*

Vostok ice core graph *http://peartrees.net/vostok-temp-vs-co2.gif*

CO_2 Time Series Graph. NOAA/PMEL *https://www.pmel.noaa.gov/co2/file/Hawaii+Carbon+Dioxide+Time-Series*

United Nations Climate Change *https://www.un.org/en/climatechange/*

CO_2 Time Series in the North Pacific Graph. Mauna Loa (ftp://aftp.cmdl.noaa.gov/products/trends/co2/co2_mm_mlo.txt) ALOHA (http://hahana.soest.hawaii.edu/hot/products/HOT_surface_CO2.txt) Ref: J.E. Dore et al, 2009. Physical and biogeochemical modulation of ocean acidification in the central North Pacific. *Proc Natl Acad Sci USA* 106: 12235-12240.

Global Change: Human Impacts on Biodiversity

Chapter

18

I. Biodiversity

A. *Biodiversity* measures the variety of all the different types of life on Earth.

B. Three Types of Diversity

1. *Genetic diversity* is the variety of different genetic traits and alleles found in the genome of a species. *Alleles* are a type of gene responsible for the variations in which a given trait can be expressed.

   i. High genetic diversity is extremely beneficial to a species.

      ➤ It confers resistance to epidemic diseases because there will likely be some individuals in a population with genetic resistance who are able to survive an otherwise fatal outbreak.

      ➤ Genetic diversity is also essential in the survival of a species during a time of ecological changes because it allows a wider range of tolerance for environmental factors among individuals of the species.

   ii. If a species loses numbers, a genetic bottleneck may occur.

      ➤ If a population has a large number of individuals, each individual has its own set of genetic alleles. A population with many individuals is likely to have more possible alleles. When a population is reduced to just a few individuals, the genetic pool of that population is now made up of only the alleles found in the individuals who are left. Even when the population grows again, the gene pool of the population is reduced because all the new members of the population are offspring of the few who survived the die-off.

➤ The African cheetah is a famous example of a genetic bottleneck. In some form, cheetahs have roamed the Earth for millions of years, but they face extinction within a few generations. Every cheetah today is almost genetically identical to every other cheetah because, at some point in the recent past, the cheetah population was reduced to a handful of individuals. Most evidence points to the end of the last ice age about 10,000 years ago, when many other large animals were driven to extinction by natural changes in the environment. Currently, there are about 10,000 cheetahs in the wild, but all are highly inbred descendants of those original few survivors.

2. *Species diversity* measures how many different species are present in an ecosystem (species richness). The relative abundance of each species (species evenness) is also taken into account. The term biodiversity usually refers to species diversity.

3. *Ecosystem diversity* is a representation of all the different types of ecosystems in an area. For example:

   i. Florida's Everglades National Park is noted for high ecosystem diversity. The park contains mangrove swamps, hardwood hammocks, cypress stands, coastal lowlands, and estuary ecosystems.

   ii. When land is cleared and converted to agricultural or urban use, ecosystem diversity is greatly reduced, which in turn damages species and genetic diversity.

C. Importance of Biodiversity

   1. Interconnectedness of Life

      i. If one or more species is removed from an ecosystem, we risk the destruction of the entire system. We rely on ecosystems to provide many of our material needs such as food, lumber, and clean water. By degrading these systems, we put our precious resources at risk.

   2. Future Resources

      i. Clearing a rainforest has been compared to burning a library without ever reading any of its books. Scientists are constantly discovering new ways to use our knowledge of other organisms to improve our lives. Many pharmaceuticals are plant-based chemicals; agronomists are using genes from rainforest plants to engineer new varieties of crop plants; and materials science engineers use our increasing knowledge of how living organisms accomplish tasks like waterproofing, climbing, and gliding to develop new technologies and materials.

3. Aesthetic and Intrinsic Value

    i.   Many believe preservation of biodiversity should not be based on how it benefits humanity. The idea is that all organisms have the right to exist on Earth. Humans do not have any special rights or ownership to use and destroy other species according to our own will.

    ii.   Religious and cultural beliefs often shape our views on biodiversity. For example, Christianity holds humans as the stewards of the Earth, whereas Hinduism teaches that humans should live in harmony with nature.

    iii.   *Ecotourism* is the practice of touring natural habitats while minimizing environmental impact.

        ➤ Central American countries such as Costa Rica and Belize make far more money by preserving their tropical ecosystems and attracting tourists for rainforest hikes, rafting trips, and tours than they would earn by cutting down the rainforest for development.

II. Endangered Species

A. A species is considered *endangered* if it is at risk of becoming extinct in the near future. Extinction occurs when the last member of a species dies. Some notable extinct species include the dodo, passenger pigeon, and the Carolina parakeet.

B. Characteristics That Make a Species Vulnerable to Extinction

1. Specialist species are so named because they have very special requirements for survival.

    i.   Feeding. The giant panda is an excellent example of a specialist feeder. Ninety-nine percent of its diet consists of bamboo. Giant pandas need access to bamboo forests in their habitat because they will not adapt to eating other foods. So, they are very dependent on a stable habitat. Changes that reduce bamboo forests have left giant pandas with a shortage of food sources; as a result, they are critically endangered.

ii. Symbiotic relationships. Many relationships between two species have become highly specialized. For example, many plants have flowers that are perfectly adapted to the beak of a species of bird or insect that pollinates them. If that relationship is disturbed, extinction could follow.

Distinguish two evolutionary strategies: (1) r-selection, for those species that produce many offspring and live in unstable environments and (2) K-selection, for those species that produce fewer offspring in stable living areas.

2. K-strategist is a species whose populations fluctuate at or near the carrying capacity (K-selected) of the environment in which they live. Many life history traits that define a species as a K-strategist also put them at risk of extinction.

    i. Large body size. Large species require more resources including food, clean water, habitat for roaming, and so on. As a result, these species tend to have small population sizes when compared to R-selected species. When a K-selected species competes with humans for these resources, there is an increased likelihood of extinction. Examples include grizzly bears, African and Indian elephants, and American bison (buffalo).

    ii. Low birth rate. K-selected species generally take several years to reach sexual maturity, have a long gestation period, and give birth infrequently (often only once a year or less) to a small number of offspring (one to three at a time). This characteristic makes them less adaptable to rapid changes in their environment. R-selected species, like E. coli bacteria, can rapidly adapt through evolution because their generation time is mere minutes. K-selected species often have generation times of twenty or thirty years.

3. Perceived as a pest species

    i. The gray wolf was near extinction in North America in the 1930s. Their natural prey is ungulates (hoofed mammals), including livestock. As a result, ranchers in the central U.S. launched an eradication campaign in which every wolf was hunted and killed.

ii.   In 1995, gray wolves were reintroduced into Yellowstone National Park under the Endangered Species Act. Today, the gray wolf population in North America is growing to the point where they may soon be removed from the endangered species list.

4.   Biomagnification harm. Top predators are prone to the effects of biomagnification. (This process is explained in Chapter 13.) But recall that toxins travel through the trophic levels of a food web, so the organisms at the top will accumulate the highest concentration of the toxin in their tissues. Sometimes this kills them outright, but more often it will disrupt their ability to reproduce, find food, or otherwise thrive in the environment.

i.   In the 1950s and 1960s, the use of the pesticide DDT caused bald eagles to accumulate this poison in their tissues as they ate fish contaminated with DDT. DDT interfered with calcium deposition in birds, which caused them to lay thin-shelled eggs. Bald eagle numbers severely declined because they were unable to lay fully calcified eggs. As a result, their eggs could not be incubated to hatching. Only since DDT has been banned in the U.S. have bald eagle populations begun to recover fully.

5.   Commercially in demand. Some species are prized for their beautiful fur or hide, medicinal purposes, or some other valuable attribute.

i.   American alligators, hawksbill sea turtles, and Sumatran tigers have been hunted for their hide, shell, or fur. These products are then made into fashion handbags, shoes, jewelry, or rugs.

ii.   Sperm whales, right whales, and fin whales were hunted freely until the mid-twentieth century. Before whaling became regulated worldwide in the 1970s, more that 1 million sperm whales were harvested. Sperm whales were especially prized because of a waxy substance called *spermaceti* that is derived from whale oil. This odorless and tasteless wax was used widely in making candles, cosmetics, lubricants, and medicating ointments. More recently, other chemical formulations have been developed with similar properties. Other species are hunted for whale oil and whale meat. Although commercial whaling is banned in the waters surrounding the U.S., other countries like Norway and Iceland continue whaling, but on a more limited basis.

6. Traditional medicines

   i. Two species of rhinoceros, the white rhino and the black rhino, are critically endangered due to habitat loss and hunting. The rhinoceros is hunted for its horn, which is prized as a raw material for carving and an ingredient in Chinese traditional medicines.

   ii. Other animals are aggressively hunted or *poached* (illegally hunted) due to the demand for their body parts for use in traditional Chinese medicine. These include all species of tigers (for their bones), Asiatic black bear (for its bile), musk deer (musk), and more than thirty-five species of seahorse (their whole body is used).

C. Causes of Extinction

1. Any event that alters the natural environment faster than its inhabitants can adapt is a risk factor for extinction. Different species will respond to risk factors in different ways, and not all will face extinction. Some species may be able to adapt quickly to environmental change or move to a new location. Here are a few ways that humans cause extinction.

2. Habitat loss is the leading cause of species extinction.

   i. As human populations continue to grow, we are encroaching more and more into formerly undeveloped areas. We are converting more land for agriculture to feed our growing population, and urbanization is increasing on a global scale as well.

   ii. Fragmentation of habitats due to highways, power line corridors, and so on, reduces area for home ranges of large roaming animals. This in turn reduces gene flow within a species, creating genetically distinct subpopulations.

   iii. Pollution is another cause of habitat loss. If we release pollution that acidifies a lake beyond the range of tolerance for its inhabitants, we have essentially destroyed their habitat, putting them at risk for extinction.

3. Hunting/Poaching

   i. Humans have hunted many species to the edge of extinction. Humans continue to hunt for meat, for vanity products like combs made of hawksbill sea turtle shell, or (in the case of American bison) merely for sport. At a time when the Earth's resources seemed limitless, it never occurred to people that extinction was possible. We see now that it is not only possible, but a real threat to many species.

4. Competition from Invasive Non-Native Species

    i.   An exotic species is one that has been taken from its natural habitat and transplanted to a new place. Sometimes this happens naturally through a storm event or via seed dispersal by birds and bats, but more often it is a result of human activities. Whether intentional or unintentional, the consequences can be devastating for native species.

    ii.   Exotic species in a new habitat often outcompete native species with similar niches because the exotic species no longer has natural predators, diseases, or other ecological checks when it is removed from its natural system.

    iii.   Bilge water transport. When a fully loaded cargo ship is unloaded, the ship rises in the water as it becomes lighter. This makes the ship unstable; before it leaves port to sail homeward, the ship will fill large tanks in its hull with ballast, or bilge water. This adds weight to the ship and lowers it in the water, increasing its stability. When it returns to its home port, the ship releases the bilge water from its tanks. In this way, thousands of exotic species have been transported all over the world to be released in ports thousands of miles from their native habitat. The introduction of these native aquatic species has caused major problems worldwide, especially in the U.S.-Canadian Great Lakes.

    iv.   Biological control gone wrong. Species are often imported for a specific purpose. Generally, we think that using natural relationships among species (biological control) to serve our purposes is a more sustainable alternative to less natural means, and often it is. But not always.

        ➤   In 1935, cane toads were introduced into northern Australia to eliminate sugar cane beetles without the need for pesticides. Cane toads have become one of the most aggressive, invasive exotic species known. They are voracious, breed rapidly, and produce a powerful toxin that is often lethal to any animal that tries to eat them. The cane toad has been a particular threat to many small marsupials, which are either eaten by the toads or killed by their toxins as a result of contact with them.

    v.   Ornamental species gone wild. Plants are imported with beautiful flowers, foliage, or pleasing scents to be used as an ornamental species in landscaping. Unfortunately, some of our most invasive exotic pests are ornamental plants that have escaped into wild ecosystems. Examples include Brazilian pepper, heavenly bamboo, and water hyacinth.

vi. Unwanted or escaped pets. Exotic pets often grow too large or prove to be too dangerous for an owner to keep. A common remedy is to free the pet by abandonment outdoors. If the climate is right, the former may become established as part of an exotic population. Throughout the southern U.S., wild populations of parrots, iguanas, and pythons exist. Most of these animals were released by irresponsible owners or escaped from their captive homes.

D. Efforts to Protect Endangered Species

1. Habitat Protection and Restoration

   i. The most effective method of protecting endangered species is to preserve their habitat. The benefit is that we also make life possible for all the other species that share that habitat.

   ii. Examples of habitat protection are designating an area as a park or wilderness area, building wildlife corridors between habitat islands, and reducing pollutants from entering the habitat through regulation of local industry and citizens.

   iii. Habitat corridors link areas of habitat that have been fragmented and allow for animals to move and maintain viable populations.

2. Captive breeding programs

   i. Some species have become so close to extinction or have so little habitat left, that just managing their wild populations may not be enough.

   ii. African cheetah, giant pandas, and some species of tigers are among the critically endangered species being bred in captivity in an attempt to diversify their gene pool and provide a refuge while new habitats are being set aside and recovered.

E. The Anthropocene

1. The anthropocene refers to the current geologic time period in which humans have been the dominant influence on climate and environmental change.

2. Scientists have speculated that this period can be characterized as one of mass extinction of many species.

III. **Legislation to Preserve Biodiversity**

A. Endangered Species Act (ESA)

1. The comprehensive Endangered Species Act (1973) provides protection for any species that is determined to be threatened or endangered with extinction. Provisions include strict enforcement of habitat protection, a ban on any activity that will disturb or endanger the life of a listed species, and a ban on the import or export of any individual organisms or product derived from an endangered species.

B. Convention on International Trade in Endangered Species (CITES)

1. This international agreement regulates the transport of living organisms or products made from any organism on the so-called red list as being endangered. Its goal is to ensure that any international trade does not contribute to the extinction of any animal or plant species.

REFERENCES

Convention on International Trade in Endangered Species of Wild Fauna and Flora: What is CITES? *http://www.cites.org/eng/disc/what.shtml*

PART IV
ENVIRONMENTAL OVERVIEW

Relating Economics and Environmental Issues

19

Although few topics in this chapter relate directly to questions on the exam, there is definitely a strong undercurrent throughout the entire exam that ties together the economy and the environment. Test-takers have reported that many questions required common sense and did not relate specifically to any one concept they had been taught. This common-sense approach recognizes that essentially many environmental decisions are based directly on the underlying economic cost. The aim of this chapter is to get you thinking about why we make decisions that, on the surface, may seem unreasonable.

I. Cost-Benefit Analysis

A. A *cost-benefit analysis* is a way of analyzing the possible options, weighing the costs against the possible benefits, and making the decision that will ultimately bring the greatest benefit. Almost every decision we make uses the cost-benefit analysis process, whether or not we are aware of it.

1. In a business setting, the costs and benefits are usually quantified in dollar amounts. Although it is easy to quantify costs in terms of how much a pollution control device or a cleanup operation will cost, other costs, like long-term health effects resulting from exposure to pollution, are more difficult to measure.

2. Some benefits of environmental-friendly initiatives are difficult to quantify. Earned profit is easy to quantify, but how much is clean air and water worth? What is the dollar value per acre of high biodiversity? Until we find standardized ways to quantify environmental costs and benefits, they will not be considered in the final equations used by corporate decision makers.

B. Externalized Costs

1. *Externalized costs* are the negative effects of an economic transaction that are not reflected in the actual cost of a product or service. When you buy an inexpensive pair of cotton pants from a store, you are not paying the full cost of the product. Here are just a few of the many externalized costs rarely considered as we are buying those pants:

i. Pollution caused by pesticides in the cotton fields, and eutrophication from fertilizer runoff into local waterways where cotton is produced.

ii. Human suffering resulting from deplorable factory working conditions and factory pollutants released in countries with little or no environmental regulation.

iii. The fossil fuels used to power farming equipment on the cotton farms, to transport the raw materials to the factory, and to transport the finished product to the consumer. Often these inexpensive products have traveled all over the globe by the time they reach your local store.

C. Full Cost Pricing

1. *Full cost pricing* is the economic approach of including the environmental and human costs in the price of goods and services.

D. Environmental Legislation

1. *Environmental legislation* is one step in the right direction. By imposing fees and fines for pollution and environmental degradation, corporations are forced to add these factors into the cost-benefit equation.

E. Economics of Prevention vs. Recovery

1. Another aspect to long- versus short-term thinking relates to the difficulty of cleaning up a mess versus the cost of avoiding the mess in the first place.

2. Pollution prevention is less costly in the long run than pollution cleanup. This is compounded by the fact that, in most cases, even after vast amounts have been spent to remediate a polluted area, it still will not be as healthy or productive as it was prior to the release of the pollutant.

3. In the past, many corporations were not held financially responsible for pollutants they released into the environment, so they did not design their factories and power plants to include pollution reduction measures.

4. Since the flurry of national and international legislation starting in the 1960s, corporations have been forced to spend millions, even billions, of dollars fixing environmental problems that their pollution has caused. There are now fines for emissions and effluent leaks, which cut into profits. This has led to businesses realizing that reducing pollution at the source is the best economic decision.

5. We know that pollution does not stay near its point of origin. Harmful gases released in the atmosphere rapidly mix with moving air currents and are carried around the globe.

    i.   After the Japanese Fukushima Daiichi nuclear meltdown in 2011, radiation was measured in the air in the U.S. within a few months.

    ii. Travel to the most remote, pristine tropical beach, and you might spot plastic trash floating along the shore. Much used plastic ultimately makes its way into the oceans, where it moves worldwide via ocean currents. Beaches in remote villages of Costa Rica and Nicaragua are now littered with plastic trash from all over Central America. Hawaiian beaches routinely become awash in trash from China and Japan.

Key Environmental Legislation

Many of the acts and treaties discussed in this chapter are discussed elsewhere in this book. This chapter is intended to be a quick reference to the legislation that is most likely to show up on the AP® exam. Except for the Kyoto Protocol, all of these acts and treaties have appeared in either the multiple-choice or free-response questions of all released exams. Be sure to familiarize yourself with these acts and treaties and don't spend too much time on others you may have learned along the way.

I. National Environmental Policy Act (NEPA) (1970)

A. One of the first federal environmental laws in the U.S., NEPA promoted the idea of sustainability.

   1. Environmental Impact Statement (EIS)

      i. Any major federal agency must submit an EIS for any activity that may have a harmful impact on the environment. The agency must outline all possible environmental effects, steps it has taken to avoid environmental harm, and justification for why any unavoidable harm may be necessary. All EISs are reviewed and revised, and must be approved before any proposed project may proceed.

   2. Environmental Protection Agency (EPA)

      i. Soon after NEPA was enacted, the United States Environmental Protection Agency was created to enforce its provisions. Its primary purpose is to protect human health and the environment and enforce standards under various state and local environmental laws.

II. Water Quality

A. In 1969, the contaminated Cuyahoga River in Cleveland, Ohio, caught fire and ignited national outrage at the level of pollution in the nation's waterways. Soon after, pressure was placed on government leaders to pass binding legislation to reduce water pollution.

  1. Clean Water Act (CWA) of 1972

    i. The mission of the Clean Water Act is to restore and maintain the chemical, physical, and biological integrity of America's waterways to support wildlife and recreation activities. The CWA is intended to reduce and prevent both point and nonpoint sources of water pollution in surface waters. There are no provisions specific to groundwater protection.

  2. Safe Drinking Water Act (SDWA) of 1974

    i. The focus of the SDWA is to maintain the purity of any water source that may potentially be used as drinking water. This includes both surface water and groundwater.

III. Air Quality

A. The Clean Air Act

  1. By 1955, federal action was needed to reduce air pollution. Regulations were issued that in 1963 became the Clean Air Act. The law provides:

    i. Authorization and funding to study air quality in order to learn more about the presence and effects of atmospheric pollutants.

    ii. Setting enforceable regulations to limit emissions from stationary sources (factories and power plants) as well as mobile sources (cars, trucks, and ships).

    iii. Developing programs to monitor and reduce acid deposition and the primary pollutants that cause its formation.

    iv. Establishing a program to phase out the use of chemicals that deplete stratospheric ozone.

    v. Establishing a cap and trade program for SO_2:

      ➤ *Cap and trade* is a system in which maximum allowable emissions are set for each industry.

➤ Businesses that can reduce their emissions below the standards are awarded credits. They can then sell their credits to other businesses that cannot meet the limits.

➤ This makes economic sense because it creates a huge financial incentive for industry to reduce emissions quickly. Allowing business firms to sell emissions credits means that many groups can rapidly recover the costs associated with reducing emissions.

B. Montreal Protocol (1987)

1. The Montreal Protocol is proof that global environmental problems can be solved by a combination of international cooperation and scientific advances. Participating nations are required to phase out the use of ozone-depleting chemicals in favor of less harmful alternatives.

C. Kyoto Protocol

1. The Kyoto Protocol (1997) adopted measures for reducing greenhouse gas emissions. When the U.S. refused to sign the protocol, the whole movement lost support.

The Clean Air Act has been the only air-quality law tested on the AP® exam. The Montreal and Kyoto Protocols are also significant because each deals with a major environmental concern.

IV. Solid and Hazardous Wastes

A. Comprehensive Environmental Response, Compensation, and Reliability Act (CERCLA) (1980)

1. CERCLA or the Superfund Act was enacted to handle industrial contamination in sites where no direct individual or party could be held responsible for cleanup.

2. The law imposed a tax on the chemical and petroleum industries and authorized the federal government to respond to the release of hazardous substances that might endanger public health or the environment.

3. The act also created a trust fund for cleaning hazardous waste sites when no responsible party could be identified.

B. Resource Conservation and Recovery Act (RCRA) (1976)

1. Commonly called the "Cradle-to-Grave Act," this legislation sets specific regulations concerning the manufacture, transport, storage, use, and ultimate disposal of a host of hazardous chemicals.

2. Its major provision requires extensive documentation at every step to ensure that hazardous wastes are disposed of properly.

V. Endangered Species Protection

A. The Endangered Species Act (ESA) (1973) is a far-reaching law that protects any species threatened or endangered with extinction (16 USC Ch. 35).

B. The ESA provides strict enforcement of habitat protection, a ban on any activity that disturbs or endangers the life of a listed species, and a ban on the import or export of any individual organism or product derived from an endangered species.

C. The Convention on Trade in Endangered Species (CITES) is an international law designed to prevent species from becoming endangered or extinct due to international trade of plants and animals. Over 180 countries and the European Union implement CITES, which accords varying degrees of protection to over 35,000 species of animals and plants.

VI. Delaney Clause

A. The Delaney Clause is part of the Food Additives Amendment (1958) that prohibits the use of a food additive if shown to cause cancer when eaten by humans or animals.

B. In 1958, the Food Additives Amendment was included in the Food, Drug, and Cosmetic Act (21 USC Ch. 9).

PART V

TEST-TAKING STRATEGIES AND PRACTICE

Strategies for the Multiple-Choice Questions

Before opening your AP® test booklet on test day, you should have your plan of action in place. Working your way through the 80 questions on the multiple-choice section may seem grueling, but if you have a strategy, you will be able to keep your momentum going.

On test day, you need, above all, to maintain a positive attitude. If you are well rested and feel prepared, you will be in an ideal position to succeed. To maintain that positive attitude, remember that you *do not need a perfect score to earn a 5* on this exam. In fact, if your goal is to pass with just a 3, you only need to earn about 50 to 55 percent of the total points on the exam. If your goal is a 4 or a 5, then aim for 60 to 70 percent of the total points. If you miss a third of the questions, you'll earn a 4 not a 5. Remember to bring a calculator (you can use it on both the multiple-choice and the free-response sections), pencils, and a blue or black ink pen. More specifically, a four-function (with square root), scientific, or graphing calculator may be used.

I. Text Analysis

You will have 90 minutes to answer 80 multiple-choice questions. That allows you less than one minute to answer a question. Bring a calculator and an easy-to-read watch with you on exam day (smartwatches are not allowed). Begin by reading through the exam questions and answering any questions you are sure you know. Circle any questions that will require a little more thought and come back to them later. Remember, you are collecting points toward your goal. As long as you answer about half or more of the questions correctly on this first pass, you will have accumulated enough points to earn a 3 without even tackling the difficult questions. Now, check your time. You should have 30 minutes or more to go back and try the harder questions.

Science Practice 3, Text Analysis is only assessed in the multiple-choice section of the exam. These questions will provide a short reading passage and ask you to identify an author's claim and the evidence cited in the article to support that claim.

II. To Guess or Not to Guess?

There is no guessing penalty on AP® exams. So make sure you don't leave any questions unanswered. On your second pass through the questions, eliminate answer choices that you know are incorrect, and then guess (if necessary) from the remaining choices. For every incorrect answer choice you can eliminate, you greatly increase your chances of guessing right. Be sure to read the whole question carefully and pay attention to the wording of each.

III. Types of Multiple-Choice Questions

The multiple-choice questions are divided into two sections, Part A and Part B.

Part A (usually the first 10 to 15 questions)

Common stimuli used in a question set might include:

➤ A world map with significant geologic features.

➤ A model of the carbon cycle.

➤ A diagram of the process of sewage treatment.

➤ A list of energy terms (e.g., wind, solar, coal, geothermal).

Part B makes up the rest of the multiple-choice questions. These questions are all similar in that they have a statement or a question followed by four answer choices.

CALCULATION QUESTIONS

There will be about five calculation questions in the multiple-choice section. These will be relatively straightforward calculations without many steps. You should know how to perform a percent change problem, apply the rule of 70 to calculate growth rate, calculate population growth rate, half-life, and use dimensional analysis for various problems.

DATA ANALYSIS QUESTIONS

The data analysis questions involve interpreting tables, graphs, and charts and looking for patterns and trends in the data. You will also be required to describe relationships among the variables. There may be 2 or 3 multiple-choice questions that relate to the same table, graph, or chart; after analyzing the data, you may be asked to explain how that relates to a broader environmental issue. For example, if you were looking at data on algal growth and temperature, you might have to explain *how* increased temperature leads to increased algal growth.

IV. Trust Your Instincts

Don't lose your confidence and start second-guessing yourself. If you can't remember the answer to a question or two, do NOT go back and start changing your answers. It is a good idea, however, to review your test to check that you have not made any careless errors while filling out the answer sheet. Do not change any answers unless you are absolutely certain that your original choice was clearly a mistake. Research shows that your first instinct is more likely to be correct; not your second guess. So relax, do your best, and don't worry about the rest!

Practice Multiple-Choice Questions

Practice with the following AP®-style questions. Then go online to access our timed, full-length practice exam at *www.rea.com/studycenter*.

1. Which of the following is the greatest contributor to depletion of stratospheric ozone?

    (A) Increased concentration of greenhouse gases from burning fossil fuels

    (B) Large-scale burning of tropical rainforests

    (C) Mercury emitted during the combustion of coal

    (D) CFCs released from refrigerants and insulation

2. Which of the following statements describes the role of carbon dioxide in climate change?

    (A) Carbon dioxide has the highest global warming potential of all greenhouse gases.

    (B) Carbon dioxide is not a strong contributor to climate change because it is taken in by plants during photosynthesis.

    (C) Combustion of fossil fuels has increased the concentration of carbon dioxide in the atmosphere, leading to increased heat trapping.

    (D) Because it has no anthropogenic sources, carbon dioxide does not influence climate change.

3. The diagram below represents a simplified model of the nitrogen cycle. Which letter represents the process of nitrogen fixation?

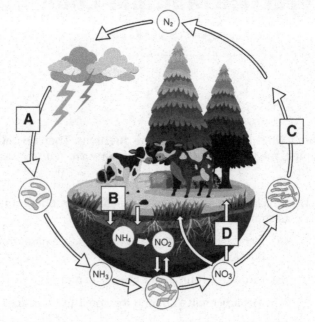

(A) A

(B) B

(C) C

(D) D

4. What organisms are responsible for the nitrogen conversions required for the process of nitrogen fixation?

(A) Algae

(B) Microbes

(C) Lichens

(D) Vectors

Latitude	Number of Mammals	Number of Reptiles	Number of Amphibians
75 degrees north	11	0	0
50 degrees north	44	5	6
25 degrees north	70	10	7
0 degrees	198	28	40
25 degrees south	99	11	33
50 degrees south	42	4	1
75 degrees south	1	0	0

The table below shows data on species richness collected at different latitudes. Reference the table for Questions 5–7.

5. Which of the following is a correct claim based on the evidence in the table?

    (A) Reptiles are the most abundant type of species found at the equator.

    (B) The northern hemisphere has greater species richness than the southern hemisphere.

    (C) There is greatest species richness found at the equator.

    (D) There is a strong correlation between increased latitude and increased species richness.

6. What approximate percent of all species at the equator are mammals?

    (A) 32.5%

    (B) 43%

    (C) 74%

    (D) 99.9%

7. Which of the following is the best explanation for the trend in the data?

(A) Mammals are the most abundant species because they are K-strategists, and amphibians and reptiles are r-strategists.

(B) Reptiles are rare at high latitudes because they are cold-blooded and need sun and warm temperatures to power their metabolism.

(C) Reptiles are rare because they have a narrow range of tolerance for pH.

(D) There are more species of mammals than amphibians and reptiles because they are larger.

The following two questions refer to the diagram below.

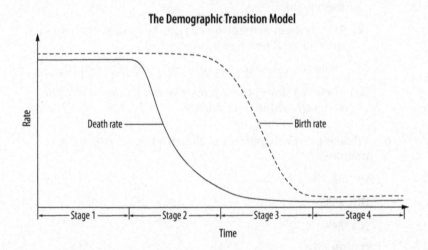

8. During which stage of the demographic transition does a country experience the most rapid growth?

(A) Stage 1

(B) Stage 2

(C) Stage 3

(D) Stage 4

9. Based on the age structure diagram for Niger, in what stage of the demographic transition does Niger belong?

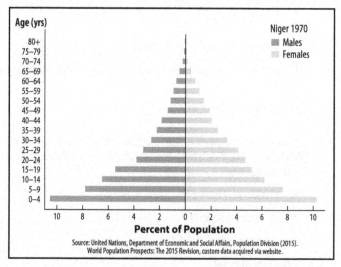

http://www.demographicdividend.org/country_highlights/niger/

(A) Stage 1

(B) Stage 2

(C) Stage 3

(D) Stage 4

10. If the average household uses eight 100-watt bulbs for 5 hours per day, how many kilowatt hours of energy do the light bulbs use in 1 year?

(A) 1,460 kWh

(B) 4,000 kWh

(C) 96,000 kWh

(D) 292,000 kWh

11. Large-scale deforestation can reduce or fragment habitat for species that require large territories. One solution to the problem of habitat fragmentation would be to

    (A) enforce CITES

    (B) implement full cost pricing

    (C) remove vulnerable species from the endangered species list

    (D) construct habitat corridors

12. Which technology captures heat from the sun to heat water and generate electricity for industrial, commercial, and residential use?

    (A) Hydrogen fuel cell

    (B) Photovoltaic cell

    (C) Solar thermal field

    (D) Geothermal

13. Which of the following statements best describes the LD50 value for a potentially toxic substance? LD50 is defined as

    (A) the dose of a substance that will kill 50% of all organisms in a population

    (B) the dose administered so that the toxin is at 50% concentration

    (C) the lowest dose of a toxin that will show a negative effect for a test population of animals

    (D) the dose that will be lethal to 50% of a tested population of organisms

Refer to Table 1 for
Questions 14 and 15.

Table 1 shows data about U.S. plastic waste management

**1960–2015 Data on Plastics in MSW by Weight
(in thousands of U.S. tons)**

Management Pathway	1960	1970	1980	1990	2000	2005	2010	2014	2015
Generation	390	2,900	6,830	17,130	25,550	29,380	31,400	33,390	34,500
Recycled	—	—	20	370	1,480	1,780	2,500	3,190	3,140
Composted	—	—	—	—	—	—	—	—	—
Combustion with Energy Recovery	—	—	140	2,980	4,120	4,330	4,530	5,010	5,350
Landfilled	390	2,900	6,670	13,780	19,950	23,270	24,370	25,190	26,010

Source: Retrieved from *https://www.epa.gov/facts-and-figures-about-materials-
waste-and-recycling/plastics-material-specific-data*

14. What percent of the plastic generated was not disposed of in a landfill in 2015?

    (A) 2.5%

    (B) 4%

    (C) 9%

    (D) 25%

15. Which of the following is the most accurate claim about the overall trend of the percent of generated plastics that ended up in the landfill from 1960–2015?

    (A) The percent of landfilled plastics has been steadily declining since the beginning of recycling and combustion in 1980.

    (B) There was a rapid decline in the percent of landfilled waste from 1980–1990.

    (C) Following the introduction of recycling and combustion, there has been an overall increase in plastic going to landfills.

    (D) The percent of landfilled plastics has remained the same from 1980–2015.

16. The Mississippi River runs north to south across the entire U.S. and drains into the Gulf of Mexico. Scientists have been recording an increase in the size and distribution of algal blooms in the late summer and fall. Which of the following statements is consistent with this finding?

    (A) Increased levels of nitrates have been recorded in water samples along the Mississippi watershed.

    (B) Levels of dissolved oxygen have increased dramatically in the Gulf of Mexico.

    (C) Fish catch in the Gulf of Mexico has been increasing in response to the increased algae.

    (D) Oil refineries in the Gulf of Mexico have been discharging hydrocarbons into the water.

17. Winds blowing parallel to the shore push warmer surface water away, causing cooler water from the deeper zones to rise to the surface. What is this phenomenon known as?

    (A) El Niño

    (B) The Windward Effect

    (C) Convection Currents

    (D) Coastal Upwellings

18. Which interaction always benefits *both* species in the relationship?

    (A) Parasitism

    (B) Mutualism

    (C) Commensalism

    (D) Synergism

19. Which of the following is a measure of the rate at which energy is converted by autotrophs into chemical energy stored in biomass?

    (A) Gross primary productivity

    (B) Net primary productivity

    (C) Secondary productivity

    (D) Primary trophic level

20. Which of the following would undergo secondary succession?

(A) A forested area that was recently burned

(B) A rainforest with high biodiversity

(C) An area of new volcanic rock

(D) A lake with high levels of dissolved oxygen

21. In which of the following situations would an r-strategist have a survival advantage over a K-strategist species?

(A) A harsh environment where abiotic conditions can change drastically

(B) An environment that favors the survival of large, intelligent organisms

(C) An environment that allows the young to survive easily

(D) An environment with predictable environmental resistance

22. A country of 25 million people has a crude birth rate (CBR) of 40 and a crude death rate of 25. What is the annual percent growth rate?

(A) 1%

(B) 1.5%

(C) 25%

(D) 5%

23. How long will it take the country in the preceding question to reach a population of 100 million?

(A) 15 years

(B) 46.7 years

(C) 70 years

(D) 93.3 years

24. Which of the following types of indoor air pollution comes primarily from non-anthropogenic sources in or around the home?

(A) Lead

(B) Volatile Organic Compounds (VOCs)

(C) Radon

(D) Carbon monoxide

25. Which of the following is a powerful ozone-depleting molecule being phased out by the Montreal Protocol?

(A) CO_2

(B) Hg

(C) CFC

(D) N_2O

ANSWERS WITH EXPLANATIONS

1. (D) is the correct answer. The chlorine from a CFC breaks off and then removes an atom from a molecule of O_3, thus destroying it. (A) Burning fossil fuels does not deplete ozone; it is releases carbon dioxide. (B) Deforestation does not impact stratospheric ozone; it causes an increase in the release of carbon. (C) Mercury is emitted from burning coal but that does not deplete ozone.

2. (C) is the correct answer. Carbon dioxide is released by the burning of fossil fuels and adds carbon to the atmosphere that would not otherwise be part of the carbon cycle, leading to more warming. (A) Carbon dioxide has a global warming potential of 1; CFCs and nitrous oxide are both larger. (B) Carbon dioxide is a strong contributor because human activities put more carbon in than is taken out through photosynthesis. (D) Carbon dioxide has anthropogenic sources burning fossil fuels.

3. (A) is the correct answer. It shows the process of atmospheric nitrogen is being converted to NH_4^+. This process is nitrogen fixation. (B) shows assimilation. (C) shows nitrification. (D) shows ammonification.

4. (B) is the correct answer. Microbes are required for the chemical conversion of molecular nitrogen to NH_4^+. (A) Algae are producers. (C) Autotrophs are self-feeders. (D) Vectors spread disease.

5. (C) is the correct answer. The equator (0 degrees) has a total of 266 species. (A) Mammals are most abundant at the equator. (B) The southern hemisphere has 142 species and the northern hemisphere has 125 species. (D) There is an inverse relationship between latitude and species richness.

6. (C) is the correct answer:

    $$198 + 28 + 40 = 266$$

    $$\frac{198}{266} \times 100 = 74\%$$

7. (B) is the correct answer. Reptiles must live in warm environments. (A) Species richness is not related to reproductive strategy. (C) pH does not explain species richness. (D) Size does not determine richness.

8. (B) is the correct answer. The transitional stage—Stage 2—is when death rates drop and birth rates remain high, causing rapid population growth. (A) Stage 1 is pre-industrial: growth is flat because birth rate and death rate are both high. (C) Stage 3 is industrial and growth levels off. (D) Stage 4 is post industrial and growth declines because birth rates fall below death rates.

9. (A) is the correct answer. Niger is in the pre-industrial Stage 1, with most of its population in the pre-reproductive age group. (B) In Stage 2, the population would see a greater percent of the population in the reproductive age group. (C) In Stage 3, there would be a greater percent of the population in the post-reproductive age group. (D) Stage 4 countries have major economic and social advances resulting in reduced family size.

10. (A) is the correct answer.

$$\frac{8 \text{ bulbs}}{1 \text{ day}} \times \frac{100 \text{ watts}}{1 \text{ bulb}} \times \frac{1 \text{ kW}}{1000 \text{ watts}} \times \frac{365 \text{ days}}{1 \text{ year}} = 292 \text{ kW} \times 5 \text{ hours} = 1460 \text{ kWh}$$

The math is incorrect for answer choices (B), (C), and (D).

11. (D) is the correct answer. Habitat corridors connect areas of intact habitat. (A) CITES bans the transport and sale of endangered species. (B) Full cost pricing reduces the demand for products with negative externalities. (C) Vulnerable species should be added to the endangered species list.

12. (C) is the correct answer. A solar thermal field has large mirrors that capture energy from the sun to heat water, generate steam, turn a turbine, and produce electricity. (A) A hydrogen fuel cell converts the energy from the reaction between hydrogen and oxygen into usable energy. (B) A photovoltaic cell is composed of silicon; when the electrons become excited by photons of energy from the sun, they generate electricity directly. (D) Geothermal energy captures heat from the Earth which can be used to heat water, make steam, and turn a turbine.

13. (D) is the correct answer. LD50 is defined as the lethal dose for 50% of a test population of organisms for any tested substance. (A) is incorrect because the LD50 measure is not of an entire population. (B) The 50% refers to those tested, not the concentration of the chemical. (C) is the description of another measure of toxicity.

14. (D) is the correct answer. To calculate the percent, simply add the data from 2015 for those three categories (Recycled = 3,140; Composted = 0; Combustion = 5,350) to get 8,490. Then calculate:

$$\frac{8,490}{34,500} = 0.246 \approx 25\%$$

The distractors for this problem were worked out to account for the likely computational errors a student might make: (A) 2.5% represents a mistake in converting the decimal to a percent. (B) 4% represents a student setting up the calculation in reverse $\left(\frac{34,500}{8,490}\right)$. (C) 9% is another distractor. Remember, you are allowed to use a calculator on the exam. In any case, use your skills of estimation and your common sense to solve this problem. You should know that $8 \times 4 = 32$, so it makes sense that 8,490 represents approximately $\frac{1}{4}$ of 34,500.

15. (C) is the correct answer. The numbers in the chart do not support choices (A) and (B). (D) could be a true statement. Composting plastics could potentially lead to a significant reduction in landfilled plastic. This choice is not the best, however, because it is not an overall trend. To detect a trend focus on longer decade intervals. (C) is the best answer because although the percent of plastic going to landfills may have decreased, the amount of plastic has grown.

16. (A) is the correct answer. Increased concentration of nitrates will stimulate the growth of algae. (B) Levels of oxygen will decrease as algae die and are decomposed. (C) Fish catch will decrease as a result of decreased dissolved oxygen. (D) There is no direct relationship between oil and algal blooms.

17. (D) is the correct answer. All of the answer choices relate to global climate patterns. (A) El Niño is a climate pattern of moving warmer and cooler air and water that leads to temporary and sometimes dramatic global changes in strong winds and rainfall patterns. The windward effect (B) is a pattern of higher rainfall on the side of a mountain facing incoming winds. Convection currents (C), caused by heating from the Earth's sun, are found in the oceans and atmospheric air, and drive many of the climactic differences across the globe in temperature and precipitation. Coastal upwellings (D) are caused by a combination of surface winds and the differing densities of warmer and cooler water. As less dense, warmer water is blown along by coastal winds at the edge of a continent, cooler, denser water beneath moves in to replace it.

18. (B) is the correct answer. Mutualism describes a relationship between organisms from different species living in a close physical arrangement, in which both partners benefit. (A) Parasitism is a type of symbiosis in which one member derives food, shelter, or other benefit from another (known as the "host"), but the host is usually harmed in some way. (C) Commensalism benefits only one of the involved two organisms. (D) Synergism produces a combined effect.

19. (A) is the correct answer. Primary productivity describes the rate at which producers in an ecosystem convert energy from the sun (through photosynthesis) or from the oxidation of certain inorganic compounds (through chemosynthesis) into organic matter/biomass within that system. Gross primary productivity (GPP) represents the total amount of chemical energy produced. (B) Net primary productivity (NPP) is the remainder of that energy after the amount is used by the producers for their own metabolic processes through cellular respiration. So, NPP = GPP – autotroph cellular respiration. Remember, GPP will always be greater than NPP because NPP accounts for energy used by the producers. (C) Secondary productivity represents heterotrophs, not autotrophs. (D) The primary trophic level involves both heterotrophs and autotrophs.

20. (A) is the correct answer. The forest already has existing soil to seed new trees. (B) is a distractor; high biodiversity is unrelated to succession. (C) would undergo primary succession and (D) dissolved oxygen is unrelated to primary and secondary succession.

21. (A) is the correct answer. An r-strategist species is better equipped to survive harsh environments. (B) Large, intelligent organisms describe K-strategist species. Also, environments that (C) allow its young to survive and (D) are ones with predictable environmental resistance, by definition, favor the K-strategist.

22. (B) is the correct answer. The growth rate is determined by the following equation: CBR – CDR/10 = % growth rate. In this example $40 - 25 = \dfrac{15}{10} = 1.5\%$.

23. (D) is the correct answer. The rule of 70 shows that 70/percent growth rate = doubling time. A growth rate of 1.5 percent is $\dfrac{70}{1.5}$ = a doubling time of 46.7 years. The population will double twice from 25 million to 50 million and again from 50 million to 100 million. This is 2 doubling times or 46.7 × 2 = 93.4.

24. (C) is the correct answer. Anthropogenic means "related to or caused by human activities." Radon comes from natural sources, often from subsurface rocks that contain small amounts of decaying uranium. The natural process of uranium disintegration yields a number of radioactive products, radon being one of them. Answer choices (A), (B), and (D) are the result of human activities: Lead (from paint and old pipes), VOCs (from carpets, paints, and other construction materials), carbon monoxide (from furnaces, gas-burning appliances, and automobiles) are all pollutants derived from human causes.

25. (C) is the correct answer. CFC compounds have been shown to destroy stratospheric ozone production and were banned by the Montreal Protocol, an international agreement. (A) Carbon dioxide is a greenhouse gas emitted from burning fossil fuels; it does not destroy ozone. (B) Mercury is released from burning coal and does not deplete ozone. (D) Nitrous oxide is a greenhouse gas emitted from agricultural practices and does destroy the ozone. But the chemical was not banned by the Montreal Protocol.

Strategies for the Free-Response Questions

On exam day, you will begin with the 90-minute multiple-choice section. Then you'll have a 10- to 15-minute break before the free-response section begins. Take the opportunity to go outside if possible. Stretch your muscles, especially your shoulders and back, to release any tension that may have built up during the previous 90 minutes.

The free-response exam booklet contains the three types of free-response questions. Recall the format for the FRQs:

Question 1: Design an investigation

Question 2: Analyze an environmental problem and propose a solution

Question 3: Analyze an environmental problem and propose a solution doing calculations

Each question will be followed by several sheets of lined paper on which to write your final answer in blue or black pen.

I. Pace Yourself

You will have 70 minutes to answer the three FRQs in any order. Each question will be worth ten points and has three to five parts that relate to a common theme. Make a plan for allocating your time and stick to it. A common complaint from low scorers is that they lose track of time. The exam proctor will not announce how much time you have left, so you must keep track for yourself. Make sure to wear a noiseless watch that doesn't have internet access.

Here are proven time-management strategies that have helped AP® test-takers.

START EACH QUESTION WITH A READING AND PLANNING PERIOD.

You will have 23 minutes to answer each question.

- Read the entire question and process it before writing anything.

- Re-read it and circle the task verbs. If the question simply says to identify, you will waste valuable time going into a lengthy explanation. Underline any key vocabulary terms that relate to the specific content topics.

- During your second reading, record thoughts and ideas that allow you to answer the question.

- This should leave you 12 minutes to organize your answer into complete sentences. Label each section of the question as you are answering it.

II. General Strategies for Answering Questions

Clearly label the question number and each part of the answer. Skip a line or two between parts to make it easier for the AP® grader to find your answers. This also gives you some space in case you want to add something later.

Keep your answers short, simple, and to the point. Use complete sentences, but do not resort to unrelated "ecobabble."

If the question states, "Identify TWO economic advantages of solar power," do not start your answer by writing "Solar power has many economic advantages over other sources of energy." Instead, begin your response by identifying the advantage.

The tasks required to answer all parts of the FRQ are scaffolded, getting progressively more difficult, and often having an easily earned point at the end. Always attempt to answer every part of the FRQ. Points are awarded for correct answers with no penalty for incorrect responses. The AP® reader will examine your entire response and look for pointable phrasing or calculations.

Be specific! Avoid broad words like *many* and *various* and phrases like *certain kinds*. If there are various air pollutants, list specific examples, or you will not earn the point. If certain chemicals cause cancer, identify them. A vague answer will earn very few points. Words like *toxic, pollution, harmful,* and *unsustainable* are good starting points, but you must elaborate with specific information to earn full credit. This is your chance to demonstrate your knowledge, so put it all down in as much detail as possible.

If a question requires a specific number of examples, e.g., TWO, only the first two responses will be scored. This is a main reason for reading and planning at the start of the question. Choose your TWO strongest responses, and go with them! Any additional answers waste valuable writing time. If you change your mind while writing, simply cross out the line you wish to delete and insert the correct response.

III. Free-Response Task Verbs

Pay special attention to the way each question is phrased. Below is a list of the most common terms you might find in a question and how to address them.

DESCRIBE/PROVIDE RELEVANT CHARACTERISTICS OF A TOPIC.

Following are the three most common terms used in questions. Start by defining any terms you will be using in your answer, then write a sentence or two adding details. Be sure to include at least one or more examples to back up your description.

EXPLAIN

You may be asked to explain *how*—which requires analyzing a relationship, pattern, or process. Explaining *why* means the reason for the pattern or process.

JUSTIFY

Justify means that you are to provide evidence to support a claim.

IDENTIFY

Simply name one or more items that fit the description—no further explanation needed. Often a question will ask you to first identify specific information about a topic and then ask you to describe its impact in a different part of the question.

MAKE A CLAIM

Write a sentence that states your conclusion from the evidence or data.

PROPOSE A SOLUTION

After you have identified a problem, you may be asked to offer a solution.

CALCULATE

You will be required to show the correct mathematic process to arrive at a correct answer. This will require showing all work, with units in the setup, for full credit.

You will be asked to analyze an environmental problem and provide a solution in 2 of the 3 FRQs! During your review, think about all the ways humans could have an impact on the environment and a possible solution. Think about the following:

- *legislation that you could pass or enforce (Clean Water Act)*

- *a practice you could encourage or provide an incentive for (construct a habitat corridor)*

- *pollution prevention or remediation technology (catalytic converters)*

IV. Types of Free-Response Questions

DESIGN AN INVESTIGATION

The first FRQ will describe an environmental scenario including a model or visual of data. You will be asked to form a research question and describe your experimental design. The AP® reader will be looking for:

- a control group and an experimental group

- the experimental group should have 4 or 5 different test levels

- dependent and independent variables

- multiple trials or repetitions

ANALYZE AN ENVIRONMENTAL PROBLEM AND PROPOSE A SOLUTION

Question 2 will require you to describe environmental processes and concepts. This could relate to everything from disposal of e-waste to

Notes

problems associated with nutrient run-off. As described above, the task verb JUSTIFY will require you to explain not only how your solution works but also how it actually solves the problem.

ANALYZE AN ENVIRONMENTAL PROBLEM AND PROPOSE A SOLUTION USING CALCULATIONS

Question 3 is similar to question 2. However, 6 of the 10 points will be earned through a series of calculations. Remember, you can use a calculator, but no credit will be earned if you do not show the setup in your exam book. Be sure to include all units in your setup and your answer.

If you have a math question that requires the answer from part (a) to do the calculation in part (b), you must use the answer you calculated in part (a), or no points will be awarded. If that answer is incorrect and you use it correctly in part (b), full credit can be earned for part (b).

In many cases, the calculation-based question requires calculations only on the first few parts. The remaining parts of the question are often easier. If you have no idea how to do the calculations, be sure to skip down and do the later parts of the question that do not require using the numbers. In many cases you can get four or five of the ten points without even attempting the calculations!

Test Tip

Actual FRQs from the AP® exam can be found on the College Board's AP® Central website. Be sure to review these questions for added practice.